T0239009

Resource Allocation in Backscatter-Assisted Communication Networks

Xiaozheng Gao · Kai Yang · Dusit Niyato · Shimin Gong

Resource Allocation in Backscatter-Assisted Communication Networks

 Springer

Xiaozheng Gao
Beijing Institute of Technology
Beijing, China

Dusit Niyato
Nanyang Technological University
Singapore, Singapore

Kai Yang
Beijing Institute of Technology
Beijing, China

Shimin Gong
Sun Yat-sen University
Guangzhou, China

ISBN 978-981-16-5126-7   ISBN 978-981-16-5127-4   (eBook)
https://doi.org/10.1007/978-981-16-5127-4

This Springer imprint is published by the registered company Springer Nature Singapore Pte Ltd.
The registered company address is: 152 Beach Road, #21-01/04 Gateway East, Singapore 189721, Singapore

Preface

In backscatter communications, a device can transmit data by modulating and reflecting the received signals and it need not generate carriers itself, which has attracted much attention over the past few years. With the rapid development of backscatter communications technology, integrating backscatter communications into traditional communications is becoming a promising way to improve the performance (e.g., spectrum efficiency and energy efficiency) of the networks. To fully improve the performance of backscatter-assisted communication networks, resource allocation is of special importance.

This book considers the resource allocation in two typical backscatter-assisted communication network structures, namely backscatter-assisted radio-frequency-powered (RF-powered) Cognitive Radio (CR) networks and backscatter-assisted hybrid relay networks. It is worth mentioning that the resource allocation in backscatter-assisted communication networks is more challenging than traditional communication networks, and the trade-off of the performance between backscatter communications and traditional communications needs to be carefully considered. Specifically, in backscatter-assisted RF-powered CR networks, more time period for backscatter communications reduces the amount of the harvested energy, which results in the decrease of the amount of the transmitted data in the harvest-then-transmit mode. Therefore, the time period of backscatter communications needs to be derived. In backscatter-assisted hybrid relay networks, the relays can perform in the active mode or the passive mode, and the working mode of the relays needs to be investigated to improve the network performance.

In particular, for the backscatter-assisted RF-powered CR networks, firstly, this book investigates the auction-based time scheduling schemes, which can guarantee that the Secondary Transmitters (STs) provide actual transmission demands to the controller. Secondly, the contract-based time assignment scheme is developed with the information asymmetry of harvested power of the ST. Thirdly, the joint access point and service selection scheme was developed for the bounded-rational STs. For the backscatter-assisted hybrid relay networks, this book investigates the relay mode selection and resource sharing to improve the network throughput. We believe that the developed resource allocation schemes in this book can provide useful guidance for the design of backscatter-assisted communication networks and future Internet of Things.

Beijing, China   Xiaozheng Gao
Beijing, China   Kai Yang
Singapore, Singapore   Dusit Niyato
Guangzhou, China   Shimin Gong

Contents

Acronyms

AF	Amplify-and-forward
AP	Access point
BnB	Branch and Bound
CR	Cognitive radio
D2D	Device-to-device
DF	Decode-and-forward
HAP	Hybrid access point
HTT	Harvest-then-transmit
IC	Incentive compatibility
IoT	Internet of Things
IP	Increasing preference
IR	Individual rationality
Max-DG	Max-Direct-Gain
Max-DR	Max-Direct-Rate
Max-RR	Max-Relay-Rate
Min-RF	Min-RF-Energy
MISO	Multi-input single-ouput
MRC	Maximal ratio combining
NOMA	Non-orthogonal multiple access
ODE	Ordinary differential equation
PR	Primary receiver
PS	Power-splitting
PSK	Phase shift keying
PT	Primary transmitter
QoS	Quality of services
RF	Radio frequency
RFID	Radio-frequency identification
SDP	Semidefinite programming
SG	Secondary gateway
SNR	Signal-to-noise ratio
ST	Secondary transmitter
TDMA	Time division multiple access

TS	Time-switching
UAV	Unmanned aerial vehicle
VCG	Vickrey-Clarke-Groves
WDTS	Winner determination and time scheduling

Chapter 1
Introduction

In this chapter, we first present the background regarding the resource allocation in backscatter-assisted communication networks. Then, we specifically discuss two typical structures of backscatter-assisted communication networks. Finally, the organization of this book is illustrated.

1.1 Background

Nowadays, the concept of the Internet of Things (IoT) is widely applied in many areas, such as assisted living and e-health, and numerous devices are connected to the wireless communication networks, which results in a significant increase in the demand of wireless services [1, 2]. Therefore, the spectrum efficiency and the energy efficiency of the communication networks need to be largely improved.

The backscatter communications technology, which was first proposed in 1948 [3], has been applied in various areas such as radio-frequency identification (RFID) [4]. With the development of the backscatter communication over the past few years, backscatter communication has been acknowledged as a promising technology to improve the spectrum efficiency and the energy efficiency of the communication networks [5–7]. Backscatter communication has several configurations: monostatic backscatter configuration, bistatic backscatter configuration, and ambient backscatter configuration [6, 8]. In monostatic backscatter configuration, the receiver and the carrier emitter are co-located on the same device, as shown in Fig. 1.1a. To avoid the round-trip path loss in the monostatic backscatter configuration, bistatic backscatter configuration has been developed, where the carrier transmitter and the receiver are separated, as shown in Fig. 1.1b. In the ambient backscatter configuration, as shown in Fig. 1.1c, the receiver and the carrier emitter are also separated. However, the carrier transmitters in the ambient backscatter configuration are available ambient radio-frequency (RF) sources, such as TV towers [5].

© The Author(s), under exclusive license to Springer Nature Singapore Pte Ltd. 2021
X. Gao et al., *Resource Allocation in Backscatter-Assisted Communication Networks*,
https://doi.org/10.1007/978-981-16-5127-4_1

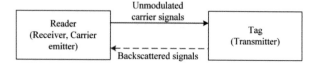

(a) Monostatic backscatter configuration

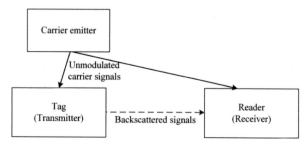

(b) Bistatic backscatter configuration

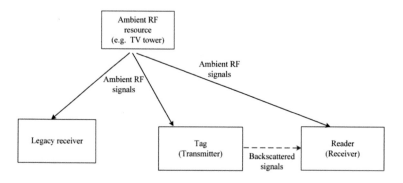

(c) Ambient backscatter configuration

Fig. 1.1 Different backscatter configurations

By using the backscatter communications technology, the device can modulate and reflect the received RF signals, such as TV signals and Wi-Fi signals, to transmit data instead of generating RF signals by itself [6]. Therefore, integrating the backscatter communications into the communication networks can bring at least three advantages as follows. First, the backscatter communications can utilize the same spectrum with the active radios, which can improve the spectrum efficiency of the network. Second, the devices can flexibly schedule their transmissions by the active mode and the backscatter mode to improve their performance. Third, the devices can serve as passive relays to backscatter active signals to enhance the performance of active transmission [9]. As such, developing backscatter-assisted communication networks has attracted much attention over the past few years, and the two typical structures of the backscatter-assisted communication networks are backscatter-assisted RF-powered cognitive radio (CR) networks and backscatter-assisted hybrid relay networks.

Backscatter-assisted RF-powered CR networks refer to the networks where the backscatter communication technology is integrated into traditional RF-powered CR networks[10, 11]. In traditional RF-powered CR networks, a typical approach for the devices to transmit data is the harvest-then-transmit (HTT) mode [12, 13]. Specifically, the devices, as the secondary transmitters (STs), harvest energy when there exist primary signals, and the STs utilize the harvested energy to transmit data when the primary channel is idle. However, when the primary channel is mostly busy, the performance of the STs is very poor due to the limited transmission time. It is noticed that the backscatter communications technology allows devices to transmit data by backscattering RF signals when the primary channel is busy. Therefore, the concept of the backscatter-assisted RF-powered CR networks has been introduced, in which the STs can choose the backscatter mode and/or the HTT mode to transmit data. Since an additional transmission mode is introduced in the backscatter-assisted RF-powered networks, the performance of the STs can be improved compared with that in the traditional RF-powered CR networks, especially when the primary channel is mostly busy [10].

Backscatter-assisted hybrid relay networks refer to the networks where the backscatter-assisted passive relay is integrated into traditional active relay communication networks [14]. Different from the active relays, the backscatter-assisted passive relays do not use any power-consuming component. When a device transmits data to the receiver actively, the other devices can serve as passive relays to assist the active transmission. Compared with traditional active relay networks, an additional passive relay mode is introduced in the backscatter-assisted hybrid relay communication networks, which improves the network throughput effectively.

Resource allocation is essential to improve the performance of wireless communication networks, especially for the backscatter-assisted communication networks. In particular, for backscatter-assisted RF-powered CR networks, when the time period for backscatter communication increases, the amount of the transmitted data in the backscatter mode increases. However, it leads to the reduction of the harvested energy of the devices, which results in the decrease of the amount of the transmitted data in the HTT mode. Therefore, how much time period for the devices to backscatter signals should be derived to improve the total amount of the transmitted data of the two modes in the backscatter-assisted RF-powered CR networks. For backscatter-assisted hybrid relay networks, whether the relays work in an active mode or a passive mode should be investigated. Therefore, the relaying strategies of the devices should be developed to improve performance of the backscatter-assisted hybrid relay networks. Motivated by the above observations, in this book, we will develop resource allocation mechanisms for backscatter-assisted communication networks, which provides useful insights for the design of backscatter-assisted communication networks and future IoT.

1.2 Network Structures

In this section, we introduce two backscatter-assisted communication network structures, namely backscatter-assisted RF-powered CR networks and backscatter-assisted hybrid relay networks.

1.2.1 Backscatter-Assisted RF-Powered CR Networks

The structures of the ST and the SG in backscatter-assisted RF-powered CR networks are shown in Fig. 1.2. Each ST is a hybrid transmitter, which contains a micro-controller, an RF energy harvester, a load modulator, an active transmitter, and each SG is a hybrid receiver, which contains an active demodulator and a backscatter demodulator. For the hybrid transmitter, the active transmitter and the load modulator are employed in the active transmission and the backscatter transmission, respectively. For the hybrid receiver, the active demodulator and the backscatter demodulator are responsible for demodulating data in the active transmission and the backscatter transmission, respectively [15]. Note that we mainly consider the overlay CR networks in this book. Therefore, in backscatter-assisted RF-powered CR networks, when the primary channel is busy, the STs can backscatter signals, as shown in Fig. 1.3, or harvest energy, as shown in Fig. 1.4. When the primary channel is idle, the STs can transmit data actively by using the previously harvested energy, as shown in Fig. 1.5. As such, the STs can transmit data in the backscatter mode and the HTT mode [10].

Specifically, in the backscatter mode, the STs employ the backscatter modulator to backscatter ambient signals to transmit data, as shown in Fig. 1.3. In particular, the STs choose the reflecting state or the non-reflecting state by switching the antenna to indicate which binary data, i.e., "1" or "0", is transmitted [5, 6]. Note that the

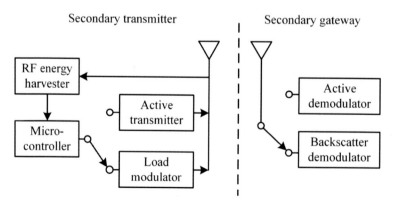

Fig. 1.2 Structures of the ST and the SG in backscatter-assisted RF-powered CR networks

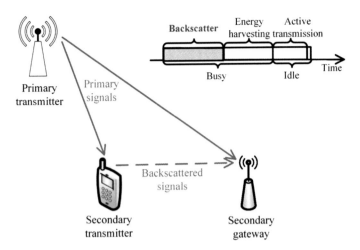

Fig. 1.3 Backscattering signals in backscatter-assisted RF-powered CR networks

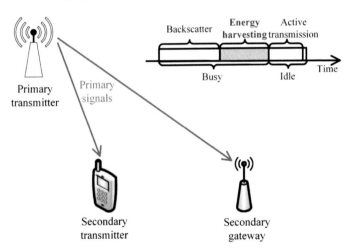

Fig. 1.4 Energy harvesting in backscatter-assisted RF-powered CR networks

backscatter transmissions will not cause any significant performance degradation to the primary receivers (PRs) when the distance between the PR and the ST is above a certain threshold [5, 10].

In the HTT mode, the ST harvests RF energy by using the RF energy harvester during the busy period, as shown in Fig. 1.4, and then utilizes the harvested energy to sustain the operations of the circuits and transmits data with active RF signals during the idle period, as shown in Fig. 1.5. Since the ST transmits data with active RF signals during the idle period only, i.e., no primary transmissions, the active transmissions would not cause any interference to the primary transmissions.

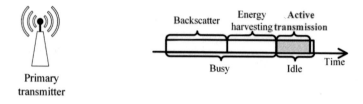

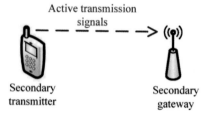

Fig. 1.5   Active transmission in backscatter-assisted RF-powered CR networks

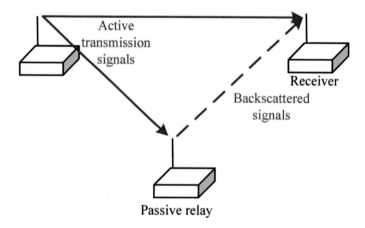

Fig. 1.6   Illustration of the relay working in the passive mode

1.2.2   Backscatter-Assisted Hybrid Relay Networks

In the backscatter-assisted hybrid relay networks, the relays can work in the active mode or the passive mode. When the relays work in the active mode, the relays can follow the decode-and-forward (DF) or the amplify-and-forward (AF) protocol to improve the network performance. When the relays work in the passive mode, they can backscatter the active RF signals to enhance the signal reception of the active transmission [14], as shown in Fig. 1.6.

Different from the active relays, where power-consuming active components such as power amplifier and oscillator are involved, the backscatter-assisted passive relays do not need the active components, which results in extremely low power consumption of passive relays. With the development of the phase shift keying (PSK) in ambient backscatter communications and the one-bit feedback-based adaptive adjustment of reflection coefficients [16], the passive relays can control the multi-path effect to enhance the reception of active transmissions. Experiment results have shown that by integrating two and three passive relays, the received signal strength can increase 111.8% and 141.7%, respectively [14], showing the effectiveness of passive relays to improve the performance of networks.

1.3 Organization

This book mainly investigates the resource allocation in backscatter-assisted communication networks. The rest of this book is organized as follows:

Chapter 2 investigates the auction-based time scheduling for backscatter-assisted RF-powered CR networks. In the networks, the STs may provide false transmission demands to the resource controller, which results in the reduction of the transmission efficiency. As such, based on auction theory, we develop time scheduling mechanisms for the networks, which can guarantee that all the STs can provide true transmission demands for the resource controller, and the transmission efficiency of the networks is therefore improved.

Chapter 3 investigates the contract-based time assignment for backscatter-assisted RF-powered CR networks. In the networks, when an SG assigns the time to the ST, the SG cannot measure the harvested power of the ST, i.e., the information of the harvested power of the ST is asymmetric. As such, based on contract theory, we develop the time assignment mechanism for the networks under the condition that the information of the harvested power is asymmetric.

Chapter 4 investigates the evolutionary game-based access point (AP) and service selection for backscatter-assisted RF-powered CR networks. In the networks, it is challenging for an ST to know the selection strategies of all the other STs. As such, based on evolutionary game, we develop the AP and service selection mechanism for the networks with bounded rationality of the STs.

Chapter 5 investigates the performance-maximized relay mode selection and resource sharing for backscatter-assisted hybrid relay networks. In the networks, the relays can work in the active mode or passive mode. To improve the network performance, whether the relays work in the active mode or the passive mode should be investigated. Also, investigating the beamforming strategies of hybrid access point (HAP) and relays can improve the network performance effectively. As such, we develop the relay mode selection and resource sharing scheme for the networks.

Chapter 6 summarizes the book and discusses the future work in the related areas.

References

1. A. Al-Fuqaha, M. Guizani, M. Mohammadi, M. Aledhari, M. Ayyash, Internet of things: a survey on enabling technologies, protocols, and applications. IEEE Commun. Surveys Tuts. **17**(4), 2347–2376 (2015)
2. V. Gazis, A survey of standards for machine-to-machine and the Internet of Things. IEEE Commun. Surveys Tuts. **19**(1), 482–511 (2017)
3. H. Stockman, Communication by means of reflected power. Proc. IRE **36**(10), 1196–1204 (1948)
4. L. Xie, Y. Yin, A.V. Vasilakos, S. Lu, Managing RFID data: challenges, opportunities and solutions. IEEE Commun. Surveys Tuts. **16**(3), 1294–1311 (2014)
5. V. Liu, A. Parks, V. Talla, S. Gollakota, D. Wetherall, J.R. Smith, Ambient backscatter: wireless communication out of thin air, in *ACM SIGCOMM* (ACM, 2013), pp. 39–50
6. N.V. Huynh, D.T. Hoang, X. Lu, D. Niyato, P. Wang, D.I. Kim, Ambient backscatter communications: a contemporary survey. IEEE Commun. Surveys Tuts. **20**(4), 2889–2922 (2018)
7. G. Yang, Y.-C. Liang, R. Zhang, Y. Pei, Modulation in the air: backscatter communication over ambient OFDM carrier. IEEE Trans. Commun. **66**(3), 1219–1233 (2018)
8. G. Wang, F. Gao, R. Fan, C. Tellambura, Ambient backscatter communication systems: detection and performance analysis. IEEE Trans. Commun. **64**(11), 4836–4846 (2016)
9. J. Xu, J. Li, S. Gong, K. Zhu, D. Niyato, Passive relaying game for wireless powered Internet of things in backscatter-aided hybrid radio networks. IEEE Internet Things J. **6**(5), 8933–8944 (2019)
10. D.T. Hoang, D. Niyato, P. Wang, D.I. Kim, Z. Han, Ambient backscatter: a new approach to improve network performance for RF-powered cognitive radio networks. IEEE Trans. Commun. **65**(9), 3659–3674 (2017)
11. D. T. Hoang, D. Niyato, P. Wang, D. I. Kim, Optimal time sharing in RF-powered backscatter cognitive radio networks, in *IEEE ICC* (IEEE, 2017), pp. 1–6
12. H. Ju, R. Zhang, Throughput maximization in wireless powered communication networks. IEEE Trans. Wireless Commun. **13**(1), 418–428 (2014)
13. Y.L. Che, L. Duan, R. Zhang, Spatial throughput maximization of wireless powered communication networks. IEEE J. Sel. Areas Commun. **33**(8), 1534–1548 (2015)
14. S. Gong, J. Xu, D. Niyato, X. Huang, Z. Han, Backscatter-aided cooperative relay communications in wireless-powered hybrid radio networks. IEEE Netw. **33**(5), 234–241 (2019)
15. X. Lu, H. Jiang, D. Niyato, D.I. Kim, Z. Han, Wireless-powered device-to-device communications with ambient backscattering: performance modeling and analysis. IEEE Trans. Wireless Commun. **17**(3), 1528–1544 (2018)
16. P.S. Yedavalli, T. Riihonen, X. Wang, J.M. Rabaey, Far-field RF wireless power transfer with blind adaptive beamforming for Internet of Things devices. IEEE Access **5**, 1743–1752 (2017)

Chapter 2
Auction-Based Time Scheduling

In this chapter, we investigate the auction-based time scheduling for backscatter-assisted RF-powered CR networks. Section 2.1 introduces the motivation of developing auction-based time scheduling mechanisms and summarizes the contributions. Section 2.2 presents the system model and the auction model. Sections 2.3 and 2.4 develop the fixed-demand and the variable-demand auction-based time scheduling mechanisms, respectively. Section 2.5 presents the simulation results, and Sect. 2.6 concludes this chapter.

2.1 Introduction

In backscatter-assisted RF-powered CR networks, there usually exist multiple STs transmitting data to the same SG [1, 2]. Therefore, the time resource of the primary channel needs to be assigned to the STs to satisfy their transmission demands. However, with more and more devices connected to the network [3, 4], the time resource becomes more congested, and it may not satisfy all the transmission demands of the STs. Since the channel conditions and quality of services (QoS) requirements are different for STs, the STs may value the time resource differently, and the SG is more likely to assign the time resource to the STs with higher valuations in their transmissions, which bring more social welfare to the network [5, 6]. Therefore, auction is expected to be an effective and fair method to the time resource assignment [5–7].

Motivated by the above observations, we take the transmission demands of STs into consideration and develop two auction-based time scheduling mechanisms for a backscatter-assisted RF-powered CR network. Specifically, according to a variety of demand requirements from STs, we develop the fixed-demand and the variable-demand auction-based time scheduling mechanisms. The major contributions of this chapter are summarized as follows.

© The Author(s), under exclusive license to Springer Nature Singapore Pte Ltd. 2021
X. Gao et al., *Resource Allocation in Backscatter-Assisted Communication Networks*,
https://doi.org/10.1007/978-981-16-5127-4_2

1. We develop an auction model for a backscatter-assisted RF-powered CR network, where the STs act as buyers, and the SG acts as the seller as well as the auctioneer. In the auction model, each ST submits its transmission demand and its unit data valuation, i.e., the valuation of transmitting unit data, to the SG to compete for the time resource. The SG is responsible for collecting the bids, performing the winner determination and time scheduling (WDTS), and calculating the payments for winner STs.

2. We develop a heuristic fixed-demand time scheduling mechanism and an optimal variable-demand time scheduling mechanism for STs. Both the WDTS strategy and the pricing scheme are developed for the two mechanisms. Specifically, in the fixed-demand case, we heuristically solve the formulated WDTS problem, which is a mixed-integer problem, and calculate the payments for winner STs based on the critical valuation. In the variable-demand case, we optimally solve the formulated WDTS problem, which is convex, and calculate the payments for STs using the generalized Vickrey-Clarke-Groves (VCG) pricing scheme.

3. We analytically evaluate the economic properties and the computational complexities for our developed two mechanisms. We prove that both two time scheduling mechanisms are individually rational and truthful. Also, we show that both our developed mechanisms are computationally efficient, i.e., the solutions can be obtained in polynomial time.

4. We evaluate and compare the performance between our developed heuristic fixed-demand and optimal variable-demand time scheduling mechanisms by extensive simulations. Also, we find that the computational complexity in the heuristic fixed-demand time scheduling mechanism is only a small fraction of that in the optimal fixed-demand time scheduling mechanism. Nonetheless, the average social welfare gap between these two mechanisms is very minor. Furthermore, we evaluate the impact of network parameters on the performance, which provides useful guidance for the time scheduling in backscatter-assisted RF-powered CR networks.

2.2 System Model and Auction Model

In this section, we first present the system model of the backscatter-assisted RF-powered CR network, and then develop an auction model to schedule the transmission time for STs.

2.2.1 System Model

We consider a time window for a backscatter-assisted RF-powered overlay CR network consisting of one PT, one SG, and K STs, as shown in Fig. 2.1. The PT, the SG, and K STs are all equipped with a single omnidirectional antenna. Note that

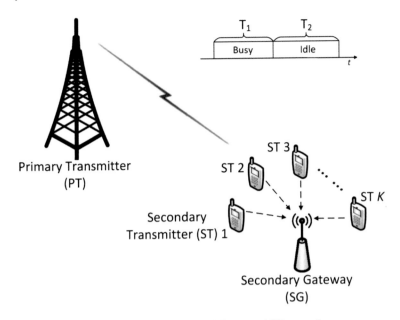

Fig. 2.1 System model of the backscatter-assisted RF-powered CR network

the auction-based time scheduling mechanisms for the multi-antenna scenario can be also developed based on the main idea of this chapter. In the primary network, the channel state is divided into the busy and the idle periods depending on whether the PT emits signals or not, and the time lengths of the busy and the idle periods are T_1 and T_2, respectively. In the secondary network, K STs, the index set of which is denoted as $\mathcal{K} = \{1, 2, \ldots, K\}$, are located within a coverage of the SG, and expect to transmit data to the same SG. To avoid interference among the STs, the time division multiple access (TDMA) is employed in the secondary network. Accordingly, only one ST can transmit data at a time. We assume that the channel gains remain fixed during the time window and can be perfectly known in the network, and the synchronization among the STs and the SG can be perfectly implemented [8, 9].

In the backscatter-assisted RF-powered CR networks, the STs can transmit data in two modes [10], namely the backscatter mode [11] and the HTT mode [9]. In particular, let τ_k^b denote the time length of ST k backscattering ambient signals. Accordingly, the amount of the transmitted data of ST k in the backscatter mode is given by

$$Z_k^b = \tau_k^b R_k^b, \tag{2.1}$$

where R_k^b is the backscatter transmission rate of ST k.

When the ST does not backscatter signals, it can harvest energy. Accordingly, the harvested energy of ST k can be given by

$$E_k^h = P_k^h \left(T_1 - \tau_k^b \right). \tag{2.2}$$

Here, P_k^h is the harvested power of ST k, and it can be expressed as $P_k^h = \eta P_{pt} \Gamma_{pt,k} d_{pt,k}^{-\alpha}$, where η is the efficiency of the energy harvesting, P_{pt} is the transmit power of the PT, $\Gamma_{pt,k}$ is the constant parameter specific to the PT and ST k, α is the path loss parameter, and $d_{pt,k}$ is the distance between the PT and ST k [12].

The harvested energy is utilized for sustaining the operations of the circuit and transmitting data during the idle period. Let τ_k^t denote the time length of ST k transmitting data in the HTT mode. Accordingly, the circuit energy consumption of ST k in the HTT mode is

$$E_k^c = P_k^c \tau_k^t, \tag{2.3}$$

where P_k^c denotes the circuit power consumption of ST k transmitting data in the HTT mode. All the harvested energy except for the circuit energy consumption is used for transmitting data [10]. Therefore, the transmit power of ST k is

$$P_k^t = \frac{E_k^h - E_k^c}{\tau_k^t}. \tag{2.4}$$

Let $d_{k,sg}$ denote the distance between ST k and the SG. Accordingly, the channel gain between ST k and the SG is given by $g_{k,sg} = \Gamma_{k,sg} d_{k,sg}^{-\alpha}$, where $\Gamma_{k,sg}$ is the constant parameter specific to ST k and the SG. Therefore, the amount of the transmitted data of ST k in the HTT mode is

$$\begin{aligned}
Z_k^t &= \varsigma \tau_k^t \beta \log_2 \left(1 + \frac{g_{k,sg} P_k^t}{N_0} \right) \\
&= \varsigma \tau_k^t \beta \log_2 \left(1 - \frac{g_{k,sg} P_k^c}{N_0} + \frac{g_{k,sg} P_k^h (T_1 - \tau_k^b)}{N_0 \tau_k^t} \right),
\end{aligned} \tag{2.5}$$

where β is the bandwidth of the channel, ς is the efficiency of the data transmission, and N_0 is the power of noise.

From (2.1) and (2.5), the total amount of the transmitted data of ST k is

$$\begin{aligned}
Z_k &= Z_k^b + Z_k^t \\
&= \tau_k^b R_k^b + \varsigma \tau_k^t \beta \log_2 \left(1 - \frac{g_{k,sg} P_k^c}{N_0} + \frac{g_{k,sg} P_k^h (T_1 - \tau_k^b)}{N_0 \tau_k^t} \right).
\end{aligned} \tag{2.6}$$

2.2.2 Auction Model

We design an auction model to allocate and schedule the time resource, i.e., the backscattering time in the backscatter mode and the transmission time in the HTT mode, for STs. The STs act as the buyers to compete for the time resource. The SG acts as the seller to provide the time resource, as well as an auctioneer responsible for collecting the bids from STs, determining the winners, scheduling the backscattering time and transmission time, and calculating the payments.

The process of the auction is summarized as follows. First, the STs submit their bids to the SG. Second, the SG determines the winners and schedules the backscattering time and transmission time for the winners. Third, the SG calculates the payment for the winners. Fourth, the SG sends the WDTS and the pricing result to each ST. Finally, the winners process the payment and transmit data to the SG. From the entire auction process, we find that the interactions between the STs and the SG mainly include the auction message exchange and signaling, the payment process, and the data transmission. The related implementation issues are considered as follows. First, the auction message exchange and signaling, such as bid submission and the auction result and price announcement, can be conducted at the beginning of the considered time window. Note that the amount of the auction message exchange and signaling is very minor compared with that of the transmitted data. Second, we can adopt a centralized credit-based transaction management model, such as in [13], to process the payments. All the STs and the SG have their accounts located in a central bank entity, which can perform the credit clearance service. When one ST needs to process its payment to the SG, the ST will send the SG a digital message representing a certain value payment. Note that the digital message can be piggybacked with the data transmission, incurring no additional energy consumption. The SG will send the digital message to the central bank entity to reduce the credit from the account of the ST and add the credit to the account of the SG. Note that the credit can be exchanged with the real cash via some agencies [14]. Third, the data can be transmitted in the backscatter mode and/or the HTT mode.

We consider that all the STs and the SG belong to different entities, and the utilities of all the STs and the SG are therefore introduced [7]. We denote the data transmission demand of ST k during the considered time window as q_k. We also denote the unit data valuation, i.e., the valuation of transmitting unit data, and the payment of ST k as v_k and p_k, respectively. The utility function of ST k, defined as the difference between the transmission valuation and the payment [5], is given by

$$u_k = x_k q_k v_k - p_k, \tag{2.7}$$

where x_k is the fraction of the transmitted data of ST k in its transmission demand. Note that p_k equals zero if ST k loses, i.e., $x_k = 0$, in the auction. Also, the utility of the SG, defined as the sum of all the received payments, is given by

$$u_{sg} = \sum_{k \in \mathcal{K}} p_k. \tag{2.8}$$

Therefore, the social welfare, defined as the sum of the utilities of all the STs and the SG, can be expressed as

$$S = \sum_{k \in \mathcal{K}} x_k q_k v_k. \tag{2.9}$$

In this chapter, we consider the following two cases:

1. Fixed-demand case: Each ST is only interested in whether its entire transmission demand can be satisfied. Transmitting some part of its transmission demand is not allowed, i.e., $x_k \in \{0, 1\}$. This case is suitable for a packet transmission in that an entire packet must be transmitted at once.
2. Variable-demand case: Transmitting some part of the ST's transmission demand is allowed, i.e., $x_k \in [0, 1]$. This case is applicable for transmitting a stream of data bits, e.g., continuous sensing data transmission.

In the above two cases, the transmission demand and the unit data valuation of ST k are only known by ST k, but not known by the other STs and the SG. Therefore, each ST needs to submit its bid to the SG. Specifically, we express the bid of ST k as $B_k = \{y_k, \varphi_k\}$, where y_k is the transmission demand of ST k, and φ_k is the maximum unit data payment, i.e., the maximum payment for transmitting unit data, that ST k is willing to pay. Note that y_k and φ_k may not be equal to q_k and v_k, respectively, if ST k can increase its utility by bidding untruthfully. However, ST k cannot increase its utility by submitting its untruthful bid in our developed mechanisms, which are proved in Sects. 2.3.3 and 2.4.3, respectively. Therefore, we rewrite the bid of ST k as

$$B_k = \{q_k, v_k\}. \tag{2.10}$$

2.3   Fixed-Demand Auction-Based Time Scheduling Mechanism

In this section, we detail the fixed-demand auction-based time scheduling mechanism for the backscatter-assisted RF-powered CR network. Specifically, the WDTS strategy and the pricing scheme are presented. Moreover, the economic properties and the computational efficiency are assessed.

2.3.1   Winner Determination and Time Scheduling

We aim to maximize the social welfare of the network in the fixed-demand time scheduling mechanism, and formulate the WDTS problem as

$$\mathcal{P}_{2.1} : \max_{x, \tau_b, \tau_t} \sum_{k \in \mathcal{K}} x_k q_k v_k \tag{2.11a}$$

$$s.t. \ Z_k \geq x_k q_k, \forall k \in \mathcal{K}, \tag{2.11b}$$

$$E_k^h - E_k^c \geq 0, \forall k \in \mathcal{K}, \tag{2.11c}$$

$$\sum_{k \in \mathcal{K}} \tau_k^b \leq T_1, \tag{2.11d}$$

$$\sum_{k \in \mathcal{K}} \tau_k^t \leq T_2, \tag{2.11e}$$

$$x_k \in \{0, 1\}, \forall k \in \mathcal{K}, \tag{2.11f}$$

$$\tau_k^b \geq 0, \forall k \in \mathcal{K}, \tag{2.11g}$$

$$\tau_k^t \geq 0, \forall k \in \mathcal{K}, \tag{2.11h}$$

where $x = \{x_k\}_{k \in \mathcal{K}}$, $\tau_b = \{\tau_k^b\}_{k \in \mathcal{K}}$, and $\tau_t = \{\tau_k^t\}_{k \in \mathcal{K}}$. The constraint in (2.11b) guarantees that the data demand requirements of winners can be met. The constraint in (2.11c) guarantees that the amount of the harvested energy is no less than that of the circuit energy consumption in the HTT mode. The constraints in (2.11d) and (2.11e) are in accordance with the TDMA, which means that the total data transmission time lengths of all STs, in backscatter and HTT modes, are no more than the time lengths of busy and idle periods, respectively. In the fixed-demand time scheduling mechanism, x_k is a binary variable, and $\mathcal{P}_{2.1}$ is a mixed-integer problem, which is NP-hard and difficult to solve. The Branch and Bound (BnB) method [15, 16] can be adopted to obtain the optimal solution to $\mathcal{P}_{2.1}$. However, the computational complexity increases rapidly with the number of STs, and the mechanism obtained by the BnB method is not computationally efficient. With more devices connected to the network [3], this complexity issue is more critical. Although relaxing the integer variables into continuous ones [17, 18] can obtain a heuristic solution to $\mathcal{P}_{2.1}$, the obtained solutions to the integer variables in $\mathcal{P}_{2.1}$ are not close to binary ones. Although there are some techniques to approximate the solutions to the binary [17], important economic properties in the auction design, such as truthfulness, are very difficult or even impossible to guarantee. Therefore, we design a heuristic strategy to address the WDTS problem by solving the winner determination and the time scheduling iteratively, which is shown in Algorithm 2.1.

From $\mathcal{P}_{2.1}$, we observe that when (i) the unit data valuation, i.e., v_k, is higher, and/or (ii) the time resource efficiency is higher, i.e., the amount of the transmitted data using the same transmission period is larger, the corresponding ST will be allowed to transmit data with a higher priority. To this end, we introduce the concept of bid density. For the time bundle of ST k, i.e., $\{\tau_k^b, \tau_k^t\}$, we define its bid size as $\tau_k^b/T_1 + \tau_k^t/T_2$. According to [5, 19], the bid density is expressed as $\frac{q_k v_k}{\sqrt{\tau_k^b/T_1 + \tau_k^t/T_2}}$. From the definition of the bid density, we highlight the fact that (i) a higher unit data valuation and/or (ii) a higher time resource efficiency, will result in a higher bid density. In Algorithm 2.1, the ST with a higher bid density will be allowed to transmit data with a higher priority, which is in accordance with the aforementioned observation from $\mathcal{P}_{2.1}$. Note that there are infinite time bundles which can meet the

---

Algorithm 2.1 A Heuristic Strategy for Solving the WDTS Problem in the Fixed-Demand Time Scheduling Mechanism

---

1: Initialize the set of winners $\mathcal{W} \leftarrow \emptyset$, the set of losers $\mathcal{L} \leftarrow \emptyset$, the set of undetermined-status STs $\mathcal{U} \leftarrow \mathcal{K}$, the time scheduling result $\mathcal{T} \leftarrow \emptyset$, and the unscheduled time lengths of busy and idle periods, $T_1' \leftarrow T_1$ and $T_2' \leftarrow T_2$, respectively.

2: **for** $k \in \mathcal{K}$ **do**

3:     Solve $\mathcal{P}_{2.2}(k)$.

4:     **if** $\mathcal{P}_{2.2}(k)$ is feasible **then**

5:         Obtain an optimal solution to $\mathcal{P}_{2.2}(k)$, i.e., $\{\tau_k^b, \tau_k^t\}$.

6:     **else**

7:         $\mathcal{L} \leftarrow \mathcal{L} \cup \{k\}$ and $\mathcal{U} \leftarrow \mathcal{U} \backslash \{k\}$.

8:     **end if**

9: **end for**

10: **while** $\mathcal{U} \neq \emptyset$ **do**

11:     $k^* \leftarrow \text{argmax}_{k \in \mathcal{U}} \dfrac{q_k v_k}{\sqrt{\tau_k^b / T_1 + \tau_k^t / T_2}}$

12:     $\mathcal{W} \leftarrow \mathcal{W} \cup \{k^*\}$ and $\mathcal{U} \leftarrow \mathcal{U} \backslash \{k^*\}$.

13:     Solve $\mathcal{P}_{2.3}$.

14:     **if** $\mathcal{P}_{2.3}$ is feasible **then**

15:         Obtain an optimal solution to $\mathcal{P}_{2.3}$, i.e., $\{\tilde{\boldsymbol{\tau}}_b^*, \tilde{\boldsymbol{\tau}}_t^*\}$ with $\tilde{\boldsymbol{\tau}}_b^* = \{\tilde{\tau}_k^{b,*}\}_{k \in \mathcal{W}}$ and $\tilde{\boldsymbol{\tau}}_t^* = \{\tilde{\tau}_k^{t,*}\}_{k \in \mathcal{W}}$.

        Here, $\tilde{\tau}_k^{b,*}$ and $\tilde{\tau}_k^{t,*}$ are the optimal solutions to τ_k^b and τ_k^t in $\mathcal{P}_{2.3}$, respectively.

16:         $\mathcal{T} \leftarrow \{\tilde{\boldsymbol{\tau}}_b^*, \tilde{\boldsymbol{\tau}}_t^*\}$

17:         $T_1' \leftarrow T_1 - \sum_{k \in \mathcal{W}} \tilde{\tau}_k^{b,*}$ and $T_2' \leftarrow T_2 - \sum_{k \in \mathcal{W}} \tilde{\tau}_k^{t,*}$.

18:     **else**

19:         $\mathcal{L} \leftarrow \mathcal{L} \cup \{k^*\}$ and $\mathcal{W} \leftarrow \mathcal{W} \backslash \{k^*\}$.

20:     **end if**

21:     **if** $\mathcal{U} = \emptyset$ **then**

22:         Stop.

23:     **end if**

24:     **for** $k \in \mathcal{U}$ **do**

25:         **if** $\tau_k^b > T_1'$ or $\tau_k^t > T_2'$ **then**

26:             Solve $\mathcal{P}_{2.2}(k)$.

27:             **if** $\mathcal{P}_{2.2}(k)$ is feasible **then**

28:                 Obtain the optimal solution to $\mathcal{P}_{2.2}(k)$, i.e.,$\{\hat{\tau}_k^b, \hat{\tau}_k^t\}$.

29:                 $\tau_k^b \leftarrow \hat{\tau}_k^b$ and $\tau_k^t \leftarrow \hat{\tau}_k^t$.

30:             **else**

31:                 $\tau_k^b \leftarrow T_1'$ and $\tau_k^t \leftarrow T_2'$.

32:             **end if**

33:         **end if**

34:     **end for**

35: **end while**

---

transmission demand of one ST. Since the SG wants to achieve a higher social welfare, the SG is willing to improve the time resource efficiency as much as possible, and the time bundle with the minimum allocated fraction of the time resource, i.e., with the highest time resource efficiency, is employed to calculate the bid density. Specifically, considering the available time resource, the time bundle with the minimum allocated fraction of the time resource for ST k can be obtained through solving the following optimization problem:

$$\mathcal{P}_{2.2}(k) : \min_{\tau_k^b, \tau_k^t} \frac{\tau_k^b}{T_1} + \frac{\tau_k^t}{T_2} \tag{2.12a}$$

$$s.t. \ Z_k \geq q_k, \tag{2.12b}$$

$$E_k^h - E_k^c \geq 0, \tag{2.12c}$$

$$0 \leq \tau_k^b \leq T_1', \tag{2.12d}$$

$$0 \leq \tau_k^t \leq T_2', \tag{2.12e}$$

where T_1' and T_2' are the unscheduled time lengths of busy and idle periods, respectively. Note that Z_k is concave with respect to $\{\tau_k^b, \tau_k^t\}$, which is shown in the following lemma.

Lemma 2.1 Z_k *is concave with respect to* $\{\tau_k^b, \tau_k^t\}$.

Proof For brevity, let $l_k = \frac{g_{k,sg}}{N_0}$. Since $\log_2(1 - l_k P_k^c + l_k P_k^h \gamma_k)$ is concave with respect to γ_k [20], its perspective function, $\tau_k^t \log_2(1 - l_k P_k^c + l_k P_k^h \gamma_k / \tau_k^t)$, is concave with respect to $\{\gamma_k, \tau_k^t\}$. Note that $\tau_k^t \log_2(1 - l_k P_k^c + l_k P_k^h (T_1 - \tau_k^b)/\tau_k^t)$ is the composition of the concave function $\tau_k^t \log_2(1 - l_k P_k^c + l_k P_k^h \gamma_k / \tau_k^t)$ with the linear function $\gamma_k = T_1 - \tau_k^b$. Therefore, from (2.5), Z_k^t is concave with respect to $\{\tau_k^b, \tau_k^t\}$. Furthermore, from (2.1), Z_k^b is a linear function with respect to τ_k^b. Therefore, from (2.6), it follows that Z_k is concave with respect to $\{\tau_k^b, \tau_k^t\}$, which completes the proof. ∎

From Lemma 2.1, we find that Z_k is concave with respect to $\{\tau_k^b, \tau_k^t\}$. Accordingly, $\mathcal{P}_{2.2}(k)$ is a convex optimization problem, which can be solved by available optimization methods easily [20].

Now we detail the heuristic strategy for solving the WDTS problem. In Algorithm 2.1, after initializing the parameters, the SG calculates the time bundle with the minimum allocated fraction of time resource for each ST, as shown from Line 2 to Line 9. Note that ST k will lose in the auction if there exists no feasible solution to $\mathcal{P}_{2.2}(k)$ with $T_1' = T_1$ and $T_2' = T_2$. The reason is that using all the time resource still cannot meet its transmission demand. Next, we determine the winner and optimize the time scheduling iteratively, as shown from Line 10 to Line 35. In each iteration, we first add the ST with the maximum bid density to the set of winners, and then optimize the time scheduling for the current set of winners, which improves the time resource efficiency from the view of the current set of winners. The time scheduling for the current set of winners is optimized by

$$\mathcal{P}_{2.3} : \min_{\tilde{\tau}_{\mathrm{b}}, \tilde{\tau}_{\mathrm{t}}} \sum_{k \in \mathcal{W}} \left(\frac{\tau_k^{\mathrm{b}}}{T_1} + \frac{\tau_k^{\mathrm{t}}}{T_2} \right) \tag{2.13a}$$

$$s.t. \ Z_k \geq q_k, \forall k \in \mathcal{W}, \tag{2.13b}$$

$$E_k^{\mathrm{h}} - E_k^{\mathrm{c}} \geq 0, \forall k \in \mathcal{W}, \tag{2.13c}$$

$$\sum_{k \in \mathcal{W}} \tau_k^{\mathrm{b}} \leq T_1, \tag{2.13d}$$

$$\sum_{k \in \mathcal{W}} \tau_k^{\mathrm{t}} \leq T_2, \tag{2.13e}$$

$$\tau_k^{\mathrm{b}} \geq 0, \forall k \in \mathcal{W}, \tag{2.13f}$$

$$\tau_k^{\mathrm{t}} \geq 0, \forall k \in \mathcal{W}, \tag{2.13g}$$

where $\tilde{\tau}_{\mathrm{b}} = \{\tau_k^{\mathrm{b}}\}_{k \in \mathcal{W}}$, $\tilde{\tau}_{\mathrm{t}} = \{\tau_k^{\mathrm{t}}\}_{k \in \mathcal{W}}$, and $\mathcal{W}$ is the current set of winners. Note that when formulating $\mathcal{P}_{2.2}(k)$ and $\mathcal{P}_{2.3}$, x_k has been replaced with the defined bid density. Since the objective functions in $\mathcal{P}_{2.2}(k)$ and $\mathcal{P}_{2.3}$ are based on the denominator of the defined bid density expression, minimizations are performed in both two problems. Note that there may be no feasible solution to $\mathcal{P}_{2.3}$. If there exists a feasible solution to $\mathcal{P}_{2.3}$, the optimal solution to $\mathcal{P}_{2.3}$, i.e., $\{\tilde{\tau}_{\mathrm{b}}^*, \tilde{\tau}_{\mathrm{t}}^*\}$, is the optimal time scheduling solution for the current set of winners. If there exists no feasible solution to $\mathcal{P}_{2.3}$, we delete the newly added ST from the set of winners, and add it to the set of losers. Note that $\mathcal{P}_{2.3}$ is also a convex optimization problem since Z_k is a concave function and other involved functions in $\mathcal{P}_{2.3}$ are linear ones. After optimizing the time scheduling, the SG continues calculating the time bundles with the minimum fraction of time resource for the undetermined-status STs,[1] as shown from Line 24 to Line 34. Note that there is no need to recalculate the time bundle if the available unscheduled time resource can meet the current time bundle requirement. If there exists a feasible solution to $\mathcal{P}_{2.2}(k)$, we update $\{\tau_k^{\mathrm{b}}, \tau_k^{\mathrm{t}}\}$ by the new optimal solution of $\mathcal{P}_{2.2}(k)$. Note that there still exists the possibility of adding ST k to the set of winners if $\mathcal{P}_{2.2}(k)$ is infeasible. This is due to the fact that the SG can optimize the time scheduling for STs to improve the time resource utilization. Therefore, in this case, we set τ_k^{b} and τ_k^{t} to be T_1' and T_2', respectively. If the time resource still cannot meet the demand of ST k after optimizing the time scheduling, ST k will be marked as a loser, as shown in Line 19. After updating the time bundles of undetermined-status STs, the SG continues a new iteration to solve the winner determination and the time scheduling until the set of undetermined-status STs is empty.

We now focus on the convergence of Algorithm 2.1. From Line 12 of Algorithm 2.1, we find that there must exist one ST deleted from the set of undetermined-status STs in each iteration of the "while" loop. Therefore, Algorithm 2.1 must terminate within K iterations of the "while" loop. Note that the required number of the iterations in the "while" loop will be less than K if there exist STs which are marked as losers from Line 2 to Line 9 in Algorithm 2.1.

---

[1] An undetermined-status ST is referred to as the ST which has not been marked as a winner or loser.

Algorithm 2.2 Deriving the Critical Valuation for Winner ST k in the Fixed-Demand Time Scheduling Mechanism

---

1: Initialize the tolerance ε, the critical valuation $c_k \leftarrow 0$, and the upper and the lower bounds of the critical valuation, $c_k^u \leftarrow \rho_k$ and $c_k^l \leftarrow 0$, respectively.
2: $v_k \leftarrow 0$
3: Run Algorithm 2.1 to obtain the new set of winners $\mathcal{W}'$.
4: **if** $k \in \mathcal{W}'$ **then**
5:   Stop.
6: **end if**
7: **while** $c_k^u - c_k^l > \varepsilon$ **do**
8:   $\rho_k \leftarrow \frac{c_k^u + c_k^l}{2}$ and $v_k \leftarrow \frac{\rho_k}{q_k}$
9:   Run Algorithm 2.1 to obtain the new set of winners $\mathcal{W}'$.
10:   **if** $k \in \mathcal{W}'$ **then**
11:     $c_k^u \leftarrow \rho_k$
12:   **else**
13:     $c_k^l \leftarrow \rho_k$ and $c_k \leftarrow \rho_k$.
14:   **end if**
15: **end while**

2.3.2 Pricing Scheme

We present the pricing scheme for the winners in our developed heuristic fixed-demand time scheduling mechanism as follows. For the sake of notational simplicity, we denote the valuation of satisfying the transmission demand of ST k as ρ_k, which can be further expressed as $\rho_k = q_k v_k$. The pricing scheme is based on the critical valuation, which is defined in Definition 2.1.

Definition 2.1 (*Critical Valuation*, [21]) The critical valuation c_k is the minimum valuation to satisfy the transmission demand of ST k, i.e., if ρ_k is no less than c_k, ST k wins; otherwise, ST k loses.

We now present the fact that there exists a critical valuation for each winner ST. Without loss of generality, for a winner ST k with the valuation $\dot{\rho}_k$, we assume that it wins in the j^*-th iteration in Algorithm 2.1. If its valuation increases from $\dot{\rho}_k$ to $\ddot{\rho}_k$, it can still win within j^* iterations. Similarly, for a loser ST k with the valuation $\dot{\rho}_k$, if its valuation decreases from $\dot{\rho}_k$ to $\ddot{\rho}_k$, it still loses in the auction. Therefore, it follows that there exists a critical valuation for each winner ST.

We now design an algorithm to derive the critical valuation for winner ST k, as shown in Algorithm 2.2. The critical valuation is derived based on the bisection method. Specifically, given other STs' bids, we first adjust v_k as 0 and rerun Algorithm 2.1 to obtain the new set of winners $\mathcal{W}'$, as shown in Lines 2 and 3, respectively. If ST k wins, the algorithm terminates and the critical valuation is 0, as shown from Line 4 to Line 6. Otherwise, we adopt the bisection method to find the ε-optimal critical valuation as shown from Line 7 to Line 15. Accordingly, in our developed heuristic fixed-demand time scheduling mechanism, the payment of winner ST k is its critical valuation, i.e., $p_k = c_k, \forall k \in \mathcal{W}$.

2.3.3 Analysis of Economic Properties

The economic properties, including individual rationality (IR) and truthfulness, can stimulate the entities, including the STs and the SG, to participate in the data transmission and sustain a stable and efficient network operation [5, 7]. In particular, if the IR cannot be guaranteed, the STs may lose motivation to join the data transmission. This can lead to an inefficient or even unsustainable network operation and cause degradation of network performance. If the truthfulness cannot be guaranteed, the STs may misreport their transmission demands or unit data valuations. As a result, the mechanism can be disordered and the time scheduled can be maliciously manipulated. In the following, we analytically evaluate the IR and the truthfulness of our developed heuristic fixed-demand time scheduling mechanism.

Theorem 2.1 *The developed heuristic fixed-demand time scheduling mechanism is individually rational for all truthful STs.*

Proof If ST k loses in the developed heuristic fixed-demand auction, the utility of ST k is zero. If ST k wins in the developed heuristic fixed-demand auction, according to Definition 2.1, we have $\rho_k \geq c_k$, and the utility of ST k is given by

$$u_k = \rho_k - c_k \geq 0. \tag{2.14}$$

In summary, we have $u_k \geq 0$ for all STs. Therefore, it follows that the developed heuristic fixed-demand time scheduling mechanism is individually rational for all truthful STs, which completes the proof.   ∎

Theorem 2.2 *With the WDTS strategy and the pricing scheme developed in Sects. 2.3.1 and 2.3.2, respectively, the developed heuristic fixed-demand time scheduling mechanism is truthful.*

Proof We denote the truthful and untruthful bids of ST k as $\{q_k, v_k\}$ and $\{y_k, \varphi_k\}$ ($\{y_k, \varphi_k\} \neq \{q_k, v_k\}$), respectively, and denote the utilities of ST k bidding truthfully and untruthfully as u_k and u'_k, respectively.

Case (i): If $y_k < q_k$, the transmission demand of ST k is never satisfied, and the valuation is zero when the amount of the transmitted data is y_k. Therefore, we have $u'_k \leq 0$.

Case (ii): If $y_k \geq q_k$, the bid density of ST k bidding with $\{q_k, c'_k/q_k\}$ is no less than that of ST k bidding with $\{y_k, c'_k/y_k\}$ in the same iteration, where c'_k is the critical valuation when ST k submits its demand y_k. According to Definition 2.1, ST k bidding with $\{y_k, c'_k/y_k\}$ wins in the auction. Furthermore, the required time resource of ST k bidding with $\{q_k, c'_k/q_k\}$ is no more than that of ST k bidding with $\{y_k, c'_k/y_k\}$. Therefore, ST k bidding with $\{q_k, c'_k/q_k\}$ also wins, and we have $c'_k \geq c_k$. If ST k bidding with $\{y_k, \varphi_k\}$ loses, then we have $u'_k = 0$. If ST k bidding with $\{y_k, \varphi_k\}$ wins, ST k bidding with $\{q_k, v_k\}$ could lead to the two following possible results:

(a) When $\rho_k \geq c_k$, ST k bidding with $\{q_k, v_k\}$ wins, and we have

$$u'_k = \rho_k - c'_k \le \rho_k - c_k = u_k. \tag{2.15}$$

(b) When $\rho_k < c_k$, ST k bidding with $\{q_k, v_k\}$ loses. Since $c_k \le c'_k$, it follows that $\rho_k < c'_k$. Therefore, we have

$$u'_k = \rho_k - c'_k < 0 = u_k. \tag{2.16}$$

In summary, each ST cannot improve its utility by bidding untruthfully, which means that our developed heuristic fixed-demand time scheduling mechanism is truthful. ∎

2.3.4 Computational Efficiency

In the following, we analyze the computational efficiency of our developed heuristic fixed-demand time scheduling mechanism.

From Algorithm 2.1, we find that the computational complexity of solving the WDTS problem is mainly determined by solving $\mathcal{P}_{2.2}(k)$ and $\mathcal{P}_{2.3}$. We denote the number of STs in the set of winners of $\mathcal{P}_{2.3}$ as $|\mathcal{W}|$. According to [20, 22], the complexities of solving $\mathcal{P}_{2.3}$ and $\mathcal{P}_{2.2}(k)$ are $O\left(|\mathcal{W}|^{3.5}\right)$ and $O(1)$, respectively. Here, $O(1)$ means that the complexity does not increase with the number of STs. Since $|\mathcal{W}| \le K$, the complexity of solving $\mathcal{P}_{2.3}$ is no more than $O(K^{3.5})$. In Algorithm 2.1, $\mathcal{P}_{2.2}(k)$ and $\mathcal{P}_{2.3}$ are solved within $\frac{(K+1)K}{2}$ and K iterations, respectively. Therefore, the total computational complexities caused by solving $\mathcal{P}_{2.2}(k)$ and $\mathcal{P}_{2.3}$ in Algorithm 2.1 are no more than $O(K^2)$ and $O(K^{4.5})$, respectively. Accordingly, the computational complexity of Algorithm 2.1 is $O(K^{4.5})$, indicating that the WDTS problem can be solved in polynomial time.

From Sect. 2.3.2, we observe that the computational complexity of the pricing scheme for each winner is mainly determined by Algorithm 2.1, and the number of times for rerunning Algorithm 2.1 is determined by the tolerance ε instead of K. Therefore, the payments of winners can also be calculated in polynomial time.

In summary, both the WDTS strategy and the pricing scheme can be obtained in polynomial time. Therefore, our developed fixed-demand time scheduling mechanism is computationally efficient.

2.4 Variable-Demand Auction-Based Time Scheduling Mechanism

In Sect. 2.3, the developed fixed-demand time scheduling mechanism is applicable for a packet transmission in that an entire packet must be transmitted at once. However, in some cases, transmitting part of a packet is allowed for STs. Therefore, we develop a variable-demand auction-based time scheduling mechanism in this section. Compared with the fixed-demand time scheduling mechanism, the variable-demand time scheduling mechanism can improve the time resource utilization, and the reasons can be explained as follows. Firstly, the time resource can be underutilized if entire transmission demand of each ST cannot be fitted to the time resource under the fixed-demand time scheduling mechanism. Secondly, we observe that the amount of transmitted data of ST k is concave with respect to $\{\tau_k^b, \tau_k^t\}$. This means that more time resource for transmitting unit data is needed when the amount of the transmitted data grows. Therefore, when transmitting some part of its transmission demand is allowed for STs, a fraction of the time resource allocated to the winner STs in the fixed-demand time scheduling mechanism can be assigned to the loser ones. Hence, the total transmitted data usually increases, which results in a higher social welfare. In the following, the WDTS strategy and the pricing scheme in the variable-demand time scheduling mechanism are presented, and the economic properties and the computational efficiency are also analytically evaluated.

2.4.1 Winner Determination and Time Scheduling

Similar to the fixed-demand time scheduling mechanism, we aim to maximize the social welfare in the variable-demand time scheduling mechanism and formulate the WDTS problem as

$$\mathcal{P}_{2.4} : \max_{x, \tau_b, \tau_t} \sum_{k \in \mathcal{K}} x_k q_k v_k \tag{2.17a}$$

$$s.t. \ 0 \le x_k \le 1, \forall k \in \mathcal{K}, \tag{2.17b}$$

$$(2.11b)\text{-}(2.11e), (2.11g), \text{ and } (2.11h).$$

We find that $\mathcal{P}_{2.4}$ is a convex optimization problem, and we can easily obtain its optimal solution $\{x^*, \tau_b^*, \tau_t^*\}$, where $x^* = \{x_k^*\}_{k \in \mathcal{K}}$, $\tau_b^* = \{\tau_k^{b,*}\}_{k \in \mathcal{K}}$, and $\tau_t^* = \{\tau_k^{t,*}\}_{k \in \mathcal{K}}$. Here, x_k^* is the fraction of the transmitted data of ST k in its transmission demand after performing the WDTS strategy, and $\tau_k^{b,*}$ and $\tau_k^{t,*}$ are the scheduled time lengths of backscattering in the backscatter mode and transmitting in the HTT mode, respectively. In the variable-demand time scheduling mechanism, we define the STs with $x_k > 0$ as winners and the STs with $x_k = 0$ as losers.

2.4.2 Pricing Scheme

In the following, we develop the pricing scheme for the winners in the variable-demand time scheduling mechanism.

We adopt the generalized VCG pricing scheme. Specifically, for calculating the payment of winner ST k, we first delete ST k from the set of STs $\mathcal{K}$ and solve the WDTS problem defined as follows:

$$\mathcal{P}_{2.5}(k): \max_{\boldsymbol{x}_{-k}, \boldsymbol{\tau}^{b}_{-k}, \boldsymbol{\tau}^{t}_{-k}} \sum_{i \in \mathcal{K}\backslash\{k\}} x_i q_i v_i \tag{2.18a}$$

$$s.t. \ Z_i \geq x_i q_i, \forall i \in \mathcal{K}\backslash\{k\}, \tag{2.18b}$$

$$E_i^h - E_i^c \geq 0, \forall i \in \mathcal{K}\backslash\{k\}, \tag{2.18c}$$

$$\sum_{i \in \mathcal{K}\backslash\{k\}} \tau_i^b \leq T_1, \tag{2.18d}$$

$$\sum_{i \in \mathcal{K}\backslash\{k\}} \tau_i^t \leq T_2, \tag{2.18e}$$

$$0 \leq x_i \leq 1, \forall i \in \mathcal{K}\backslash\{k\}, \tag{2.18f}$$

$$\tau_i^b \geq 0, \forall i \in \mathcal{K}\backslash\{k\}, \tag{2.18g}$$

$$\tau_i^t \geq 0, \forall i \in \mathcal{K}\backslash\{k\}, \tag{2.18h}$$

where $\boldsymbol{x}_{-k} = \{x_i\}_{i \in \mathcal{K}\backslash\{k\}}$, $\boldsymbol{\tau}^b_{-k} = \{\tau_i^b\}_{i \in \mathcal{K}\backslash\{k\}}$, and $\boldsymbol{\tau}^t_{-k} = \{\tau_i^t\}_{i \in \mathcal{K}\backslash\{k\}}$. It is obvious that $\mathcal{P}_{2.5}(k)$ is also a convex optimization problem, and we can obtain its optimal solution $\{\bar{\boldsymbol{x}}_{-k}, \bar{\boldsymbol{\tau}}^b_{-k}, \bar{\boldsymbol{\tau}}^t_{-k}\}$, where $\bar{\boldsymbol{x}}_{-k} = \{\bar{x}_i\}_{i \in \mathcal{K}\backslash\{k\}}$, $\bar{\boldsymbol{\tau}}^b_{-k} = \{\bar{\tau}_i^b\}_{i \in \mathcal{K}\backslash\{k\}}$, and $\bar{\boldsymbol{\tau}}^t_{-k} = \{\bar{\tau}_i^t\}_{i \in \mathcal{K}\backslash\{k\}}$. The payment of ST k, which is defined as the marginal valuation, is expressed as

$$p_k = \sum_{i \in \mathcal{K}\backslash\{k\}} \bar{x}_i q_i v_i - \sum_{i \in \mathcal{K}\backslash\{k\}} x_i^* q_i v_i. \tag{2.19}$$

We now present that we have $p_k \geq 0$ for each winner ST k. Comparing the constraints in $\mathcal{P}_{2.4}$ and $\mathcal{P}_{2.5}(k)$, we observe that $\left\{\boldsymbol{x}^*_{-k}, \boldsymbol{\tau}^{b,*}_{-k}, \boldsymbol{\tau}^{t,*}_{-k}\right\}$, where $\boldsymbol{x}^*_{-k} = \{x_i^*\}_{i \in \mathcal{K}\backslash\{k\}}$, $\boldsymbol{\tau}^{b,*}_{-k} = \{\tau_i^{b,*}\}_{i \in \mathcal{K}\backslash\{k\}}$, and $\boldsymbol{\tau}^{t,*}_{-k} = \{\tau_i^{t,*}\}_{i \in \mathcal{K}\backslash\{k\}}$, is also a feasible solution to $\mathcal{P}_{2.5}(k)$. Since $\{\bar{\boldsymbol{x}}_{-k}, \bar{\boldsymbol{\tau}}^b_{-k}, \bar{\boldsymbol{\tau}}^t_{-k}\}$ is the optimal solution to $\mathcal{P}_{2.5}(k)$, we have

$$\sum_{i \in \mathcal{K}\backslash\{k\}} \bar{x}_i q_i v_i \geq \sum_{i \in \mathcal{K}\backslash\{k\}} x_i^* q_i v_i. \tag{2.20}$$

Therefore, it immediately follows that $p_k \geq 0$ for each winner ST k.

2.4.3 Analysis of Economic Properties

In the following, we assess the economic properties of our developed variable-demand time scheduling mechanism.

Theorem 2.3 *The developed variable-demand time scheduling mechanism is individually rational for all truthful STs.*

Proof If ST k loses in the developed variable-demand auction, the utility of ST k is zero.

If ST k wins in the developed variable-demand auction, the utility of ST k is

$$u_k = x_k^* q_k v_k - p_k = \sum_{i \in \mathcal{K}} x_i^* q_i v_i - \sum_{i \in \mathcal{K} \setminus \{k\}} \bar{x}_i q_i v_i. \tag{2.21}$$

We introduce some auxiliary variables $\bar{x}_k$, $\bar{\tau}_k^b$, and $\bar{\tau}_k^t$, and let $\bar{x}_k = 0$, $\bar{\tau}_k^b = 0$, and $\bar{\tau}_k^t = 0$. Therefore, the utility of ST k can be further expressed as

$$u_k = \sum_{i \in \mathcal{K}} x_i^* q_i v_i - \sum_{i \in \mathcal{K}} \bar{x}_i q_i v_i. \tag{2.22}$$

Since $\{\bar{x}_{-k}, \bar{\tau}_{-k}^b, \bar{\tau}_{-k}^t\}$ is the optimal solution to $\mathcal{P}_{2.5}(k)$, it is easy to conclude that $\{\{\bar{x}_k, \bar{x}_{-k}\}, \{\bar{\tau}_k^b, \bar{\tau}_{-k}^b\}, \{\bar{\tau}_k^t, \bar{\tau}_{-k}^t\}\}$ is a feasible solution to $\mathcal{P}_{2.4}$. Since $\{x^*, \tau_b^*, \tau_t^*\}$ is the optimal solution to $\mathcal{P}_{2.4}$, we have

$$\sum_{i \in \mathcal{K}} x_i^* q_i v_i \geq \sum_{i \in \mathcal{K}} \bar{x}_i q_i v_i. \tag{2.23}$$

Therefore, we have $u_k \geq 0$.

In summary, it follows that $u_k \geq 0$ for all truthful STs, which means that the developed variable-demand time scheduling mechanism is individually rational for all truthful STs. ∎

Theorem 2.4 *With the WDTS strategy and the pricing scheme developed in Sects. 2.4.1 and 2.4.2, respectively, the developed variable-demand time scheduling mechanism is truthful.*

Proof We denote the truthful and untruthful bids of ST k as $\{q_k, v_k\}$ and $\{y_k, \varphi_k\}$ ($\{y_k, \varphi_k\} \neq \{q_k, v_k\}$), respectively, denote the utilities of ST k bidding truthfully and untruthfully as u_k and u_k', respectively, denote the WDTS problems when ST k bids truthfully and untruthfully as $\mathcal{P}_{2.4}$ and $\mathcal{P}_{2.4}'$, respectively, and denote their solutions as $\{x^*, \tau_b^*, \tau_t^*\}$ and $\{x', \tau_b', \tau_t'\}$, respectively, where $x' = \{x_i'\}_{i \in \mathcal{K}}$. The utility of ST k submitting its truthful bid is given by (2.21).

Case (i): When $x_k' y_k \leq q_k$, the utility of ST k bidding untruthfully is given by

$$u_k' = x_k' y_k v_k + \sum_{i \in \mathcal{K} \setminus \{k\}} x_i' q_i v_i - \sum_{i \in \mathcal{K} \setminus \{k\}} \bar{x}_i q_i v_i. \tag{2.24}$$

Note that $\{\{x_k' y_k / q_k, x_{-k}'\}, \tau_b', \tau_t'\}$ is also a feasible solution to $\mathcal{P}_{2.4}$, where $x_{-k}' = \{x_i'\}_{i \in \mathcal{K} \setminus \{k\}}$. Since $\{x^*, \tau_b^*, \tau_t^*\}$ is the optimal solution to $\mathcal{P}_{2.4}$, we have

$$\sum_{i \in \mathcal{K}} x_i^* q_i v_i \geq x_k' y_k v_k + \sum_{i \in \mathcal{K} \setminus \{k\}} x_i' q_i v_i. \tag{2.25}$$

From (2.21), (2.24), and (2.25), we have $u_k \geq u_k'$.

Case (ii): When $x_k' y_k > q_k$, the valuation of the transmitted data is $q_k v_k$. The reason is that there is no additional valuation when ST k is allowed to transmit more data than its transmission demand. In this case, the utility of ST k bidding untruthfully can be expressed as

$$u_k' = q_k v_k + \sum_{i \in \mathcal{K} \setminus \{k\}} x_i' q_i v_i - \sum_{i \in \mathcal{K} \setminus \{k\}} \bar{x}_i q_i v_i. \tag{2.26}$$

Note that $\{\{1, x_{-k}'\}, \tau_b', \tau_t'\}$ is also a feasible solution to $\mathcal{P}_{2.4}$ when $x_k' y_k > q_k$. Since $\{x^*, \tau_b^*, \tau_t^*\}$ is the optimal solution to $\mathcal{P}_{2.4}$, we have

$$\sum_{i \in \mathcal{K}} x_i^* q_i v_i \geq q_k v_k + \sum_{i \in \mathcal{K} \setminus \{k\}} x_i' q_i v_i. \tag{2.27}$$

From (2.21), (2.26), and (2.27), we have $u_k \geq u_k'$.

In summary, from Cases (i) and (ii), we have $u_k \geq u_k'$, which means that our developed variable-demand time scheduling mechanism is truthful. ∎

2.4.4   Computational Efficiency

We now analyze the computational complexity of the developed variable-demand time scheduling mechanism. The computational complexity is mainly from the WDTS strategy and the pricing scheme. In the WDTS strategy, the convex optimization problem $\mathcal{P}_{2.4}$ is solved, and its computational complexity is $O(K^{3.5})$. In the pricing scheme, $\mathcal{P}_{2.5}(k)$, the computational complexity of which is also $O(K^{3.5})$, is solved within K iterations. Therefore, the computational complexity in the pricing scheme is $O(K^{4.5})$, and the total computational complexity of the developed variable-demand time scheduling mechanism is $O(K^{4.5})$, indicating that the developed variable-demand time scheduling mechanism is computationally efficient.

2.5   Simulation Results

In this section, we present the simulation results to evaluate the performance of the developed time scheduling mechanisms. In the network, the bandwidth β is 10 MHz, the time window is 100 s, the distance between the PT and the SG is 100 m, the distances between the SG and STs are within 5 m, the transmission rates of STs in

Fig. 2.2 Computational efficiency comparison between our developed heuristic and the optimal fixed-demand time scheduling mechanisms

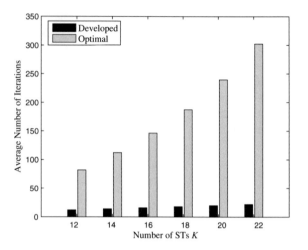

the backscatter mode are all 0.2 Mbps, and the transmission demands of STs during the time window period are all 10 Mbits. Unless otherwise specified, the unit data valuations of STs, which depend on the applications of STs, follow the uniform distribution with $v_k \in [0, 10]$/Mbits [6, 23], and the time lengths of busy and idle periods are 70 s and 30 s, respectively.

We first examine the computational efficiencies of our developed time scheduling mechanisms. In Fig. 2.2, we use the optimal fixed-demand time scheduling mechanism as a benchmark, and compare the computational efficiency between our developed heuristic fixed-demand time scheduling mechanism and the benchmark mechanism. The optimal fixed-demand time scheduling mechanism is derived by the typical BnB method in [16]. In both the heuristic and the optimal fixed-demand time scheduling mechanisms, the payments of winners are calculated by solving the WDTS problems. Therefore, we discuss the computational efficiency by comparing the complexities of solving the WDTS problems. From the analysis in Sect. 2.3.4, we find that solving $\mathcal{P}_{2.3}$ dominates the computational complexity in the heuristic fixed-demand time scheduling mechanism. In the optimal fixed-demand time scheduling mechanism by the BnB method, the relaxed problem of $\mathcal{P}_{2.1}$, the computational complexity of which is also $O(K^{3.5})$, is solved iteratively. Therefore, we consider the average numbers of iterations for solving $\mathcal{P}_{2.3}$ and the relaxed problem of $\mathcal{P}_{2.1}$ to evaluate the computational complexities of the heuristic and the optimal fixed-demand time scheduling mechanisms, respectively. We find that the average number of iterations in our developed heuristic fixed-demand time scheduling mechanism is much fewer than that in the optimal fixed-demand time scheduling mechanism. We also find that the rate of increase for the average number of iterations in our developed heuristic fixed-demand time scheduling mechanism is much smaller than that in the optimal fixed-demand time scheduling mechanism, which indicates that the developed heuristic fixed-demand time scheduling mechanism is scalable and efficient even for a large number of STs. Furthermore, in the variable-demand time

Fig. 2.3 Average social welfare comparison among different cases and scenarios

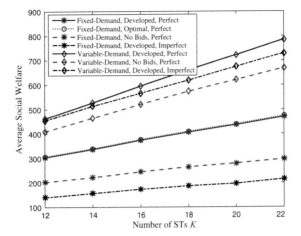

scheduling mechanism, the payments for each winner can also be calculated by solving the WDTS problem, and the computational complexity of the variable-demand time scheduling mechanism can be also evaluated by assessing the complexity of the WDTS problem in the variable-demand time scheduling mechanism. We find that the relaxed problem of $\mathcal{P}_{2.1}$ in the first iteration of the optimal fixed-demand time scheduling mechanism is the WDTS problem in the variable-demand time scheduling mechanism, (i.e., $\mathcal{P}_{2.4}$). Accordingly, the average number of iterations is always 1 for the variable-demand time scheduling mechanism, and the mechanism is also scalable and efficient for a large number of STs.

We next compare the average social welfare among different cases and scenarios in Fig. 2.3. The legend "Perfect" in this figure is in accordance with the aforementioned assumption that the network can obtain the perfect channel gains and can implement the synchronization among the STs and the SG perfectly. The legend "No Bids" in this figure means that the STs do not submit bids to the SG. In this case, the SG does not have information about the transmission demands and the unit data valuations of STs, and schedules the time resource for the STs by maximizing the throughput of the network. Note that although the SG does not take the STs' bid information into account, the transmission demands and the unit data transmission valuations of STs are still diverse and will be considered when calculating the social welfare. We observe that in the fixed-demand and the variable-demand cases, the average social welfare of our developed mechanisms is at least 45% and 10% higher than those of the mechanisms without bidding information, respectively. This results demonstrate that our developed time scheduling mechanisms are effective and can bring more social welfare to the network. Also, comparing our developed heuristic fixed-demand time scheduling mechanism with the optimal fixed-demand time scheduling mechanism, we observe that the average social welfare gap between two mechanisms is no more than 2% of the optimal social welfare, which is very minor. By contrast, the computational complexity of the developed heuristic mechanism is

Fig. 2.4 Average social welfare versus the number of STs in the developed heuristic fixed-demand time scheduling mechanism

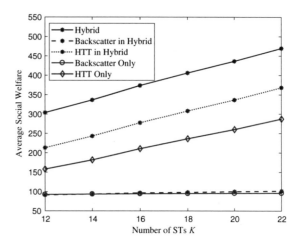

just a small fraction of that in the optimal mechanism. Furthermore, we observe that the average social welfare of our developed variable-demand time scheduling mechanism is at least 50% higher than that of the heuristic fixed-demand time scheduling mechanism. Compared with the developed fixed-demand time scheduling mechanism, the SG can assign the time resource more flexibly in the variable-demand time scheduling mechanism, and the time resource utilization is therefore improved. We further assess the average social welfare achieved by our developed time scheduling mechanisms when the time synchronization errors are within 0.1 s and the relative channel gain estimation errors are within 20%, which is marked as "Imperfect" in Fig. 2.3. From this figure, we observe that the average social welfare will be reduced due to the imperfect knowledge of the parameters. It is notable that the decrement of the average social welfare in the fixed-demand time scheduling mechanism is much larger than that in the variable-demand time scheduling mechanism. This is due to the fact that there is a jump in terms of the STs' valuations in the fixed-demand time scheduling mechanism. More specifically, in the fixed-demand time scheduling mechanism with imperfect knowledge of the parameters, the transmitted data of some winner STs may not satisfy their transmission demands, and the valuations of these STs become zero. Accordingly, the social welfare decreases significantly due to the imperfect knowledge of the parameters in the fixed-demand time scheduling mechanism.

We now evaluate the impact of the number of STs on the average social welfare and compare the average social welfare between the hybrid strategy, i.e., the backscatter-assisted RF-powered strategy, and two benchmark schemes, i.e., the backscatter only scheme and the HTT only scheme. In Figs. 2.4 and 2.5, we plot the average social welfare versus the number of STs using the developed heuristic fixed-demand time scheduling mechanism and using the developed variable-demand time scheduling mechanism, respectively. We also plot the social welfare achieved by the backscatter mode and the HTT mode, which are respectively marked as "Backscatter in Hybrid"

Fig. 2.5 Average social welfare versus the number of STs in the developed variable-demand time scheduling mechanism

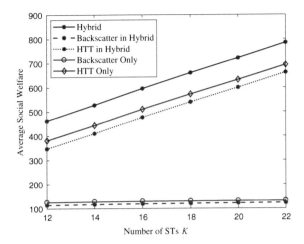

and "HTT in Hybrid", when the hybrid strategy is employed in our developed fixed-demand and variable-demand time scheduling mechanisms. From these figures, we find that both the backscatter mode and the HTT mode contribute to the transmissions between the STs and the SG, and improve the social welfare to the network. In addition, it is clear that the average social welfare increases with the number of STs in both our developed mechanisms. More STs bring more choices to the SG, and the SG can assign the time resources to the STs with higher valuations in their transmissions. As a result, more STs bring higher social welfare. Furthermore, we find that the average social welfare using the hybrid strategy outperforms that using the benchmark schemes. Compared with the backscatter only scheme, the hybrid strategy improves the idle time resource utilization. Compared with the HTT only scheme, the hybrid strategy improves the busy time resource utilization. Therefore, the hybrid strategy improves the social welfare compared with the benchmark schemes. In Fig. 2.4, we observe that the average social welfare of the hybrid strategy is larger than the sum of the average social welfare of the HTT only scheme and the backscatter only scheme. Note that the time resource can be underutilized if entire transmission demand of any loser ST cannot be fitted to the time resource under the fixed-demand time scheduling mechanism. In the hybrid strategy, each ST can use the combination of the busy time period and the idle time period. Therefore, the residual busy time period in the hybrid strategy is usually less than that in the backscatter only scheme, and the residual idle time period is usually less than that in the HTT only scheme. Accordingly, the time resource in the hybrid strategy is utilized more efficiently, and the social welfare of the hybrid strategy is larger than the sum of social welfare of the backscatter only scheme and the HTT only scheme.

We now discuss the impact of the percentage of the busy period in a certain time window, i.e., T_1/T, on the average social welfare in the developed heuristic fixed-demand time scheduling mechanism, where T is the time length of the considered time window, i.e., $T = T_1 + T_2$. Figure 2.6 plots the average social welfare versus

Fig. 2.6 Average social welfare versus the percentage of busy period in the considered time window in the developed heuristic fixed-demand time scheduling mechanism when $K = 20$

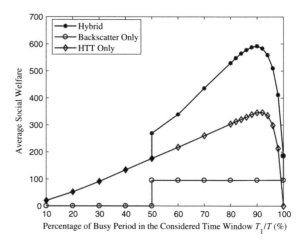

T_1/T using the developed heuristic fixed-demand time scheduling mechanism when $K = 20$. For the backscatter only scheme, there is a jump when $T_1/T = 50\%$ or 100%. Since the transmission demand for each ST is 10 Mbits, and the transmission rate in the backscatter mode is 0.2 Mbps, 50 s (i.e., $T_1/T = 50\%$) is needed to satisfy the transmission demand of one ST. When $T_1/T < 50\%$, the backscattering time cannot satisfy the transmission demand of one ST, and therefore the average social welfare is 0. When $50\% \le T_1/T < 100\%$, the ST with the highest transmission valuation is allowed to transmit data, and the social welfare is the same as the highest transmission valuation. When $T_1/T = 100\%$, the STs with the highest and the second highest transmission valuations are permitted to transmit data, and the social welfare is the sum of the highest and the second highest transmission valuations. For the HTT only scheme, we observe that the average social welfare first increases and then decreases with T_1/T. Both the amount of the harvested energy and the time length of transmitting data influence the achieved social welfare in the HTT only scheme. When T_1/T is relatively small, the harvested energy dominates the transmission capacities of STs. With the increase of T_1/T, the harvested energy becomes larger, and the transmission capacities of STs increase. Therefore, the transmission demands of more STs can be satisfied, and a higher social welfare is achieved. When T_1/T is relatively large, the transmission time dominates the transmission capacities of STs. With the increase of T_1/T, the transmission time decreases, and the transmission capacities of STs decrease, which results in a lower social welfare. For the hybrid strategy, we find that when T_1/T is relatively small, the average social welfare is the same as that in the HTT only scheme. The reason is that the STs are more likely to harvest energy in the busy period when T_1/T is relatively small. We also observe that there is a jump when $T_1/T = 50\%$, which is similar to the backscatter only scheme. Furthermore, it is clear that the average social welfare first increases and then decreases with T_1/T, which is similar to the HTT only scheme.

Fig. 2.7 Average social welfare versus the percentage of busy period in the considered time window in the developed variable-demand time scheduling mechanism when $K = 20$

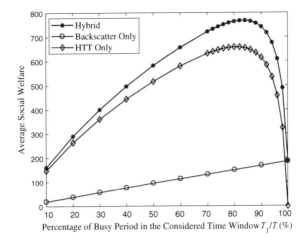

We now discuss the impact of the percentage of busy period in a certain time window on the average social welfare in the developed variable-demand time scheduling mechanism. Figure 2.7 plots the average social welfare versus T_1 / T using the developed variable-demand time scheduling mechanism when $K = 20$. For the backscatter only scheme, we observe that the average social welfare increases with T_1 / T. The reason is that the backscattering time increases with T_1 / T. For both the HTT only and the hybrid schemes, the average social welfare first increases and then decreases with T_1 / T. Comparing with the hybrid fixed-demand time scheduling mechanism, we find that there is no jump in the hybrid variable-demand time scheduling mechanism. The reason is that satisfying some part of their transmission demands is allowed for STs, and the STs can fully utilize the time resource to transmit data to the SG. Compared the result in Fig. 2.6 with that in Fig. 2.7, the developed variable-demand time scheduling mechanism outperforms the developed heuristic fixed-demand time scheduling mechanism in terms of the average social welfare, which is in accordance with the result in Fig. 2.3.

We finally discuss the impact of the unit data valuation on the winner determination results in our developed mechanisms. Figure 2.8 plots the probability of ST 1 winning in the auction versus its unit data valuation v_1 in the developed fixed-demand time scheduling mechanism when $d_{1,\text{sg}} = 3$ m and $K = 25$, 30, or 35. We find that the probability of ST 1 winning increases when its unit data valuation increases. The higher valuation leads to a higher priority to satisfy its transmission demand, and hence the ST has a higher probability to win in the auction. We also find that the probability of ST 1 winning decreases when the number of STs becomes larger. More STs result in much higher demands of the time resource, which means that it is more competitive to obtain the time resource. Therefore, when the number of STs increases, ST 1 must have a higher valuation if it wants to achieve the same probability of winning. Figure 2.9 plots the fraction of transmitted data of ST 1 in its transmission demand (i.e., x_1^*) versus its unit data valuation v_1 in the developed

Fig. 2.8 The probability of ST 1 winning versus its unit data valuation in the developed heuristic fixed-demand time scheduling mechanism when $d_{1,\text{sg}} = 3$ m and $K = 25, 30,$ or 35

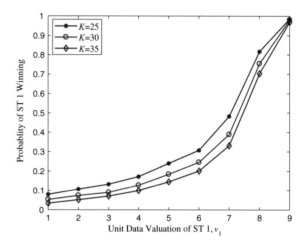

Fig. 2.9 The fraction of the transmitted data of ST 1 in its transmission demand versus its unit data valuation in the developed variable-demand time scheduling mechanism when $d_{1,\text{sg}} = 3$ m and $K = 25, 30,$ or 35

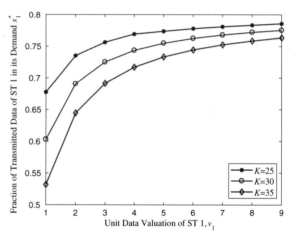

variable-demand time scheduling mechanism when $d_{1,\text{sg}} = 3$ m and $K = 25, 30,$ or 35. We find that x_1^* increases when v_1 grows. Since the SG is more likely to assign the time resource to the ST with a higher unit valuation, the transmitted data of ST 1 is increased when v_1 increases. Similar to Fig. 2.8, x_1^* decreases with the number of STs, and ST 1 must have a higher unit data valuation if it wants to transmit the same amount of data.

2.6 Conclusions

In this chapter, we have developed an auction model for backscatter-assisted RF-powered CR networks. According to a variety of demand requirements from STs, we have designed two auction-based time scheduling mechanisms for the fixed-demand and the variable-demand cases. Specifically, we have developed both the WDTS strategy and the pricing scheme for the two developed mechanisms. Accordingly, we have proved that both our developed mechanisms are individually rational, truthful, and computationally efficient. Simulations have been conducted to evaluate the performance of our developed mechanisms. The results have presented that the computational complexity in our developed heuristic fixed-demand time scheduling mechanism is only a small fraction of that in the optimal fixed-demand time scheduling mechanism, whereas the social welfare gap between the two mechanisms is very minor. The results have also shown that the hybrid backscatter-assisted RF-powered network outperforms the backscatter only or the HTT only network, and the average social welfare in the variable-demand time scheduling mechanism is larger than that in the fixed-demand time scheduling mechanism. Furthermore, the impact of some parameters, including the number of STs, the percentage of the busy period in a certain time window, and the unit data valuation, on the performance of our developed mechanisms have been evaluated, which provides useful guidance for the time scheduling in backscatter-assisted RF-powered CR networks.

References

1. D.T. Hoang, D. Niyato, P. Wang, D.I. Kim, Optimal time sharing in RF-powered backscatter cognitive radio networks, in *IEEE ICC* (IEEE, 2017), pp. 1–6
2. W. Wang, D.T. Hoang, D. Niyato, P. Wang, D.I. Kim, Stackelberg game for distributed time scheduling in RF-powered backscatter cognitive radio networks. IEEE Trans. Wireless Commun. **17**(8), 5606–5622 (2018)
3. F. Javed, M.K. Afzal, M. Sharif, B.-S. Kim, Internet of Things (IoT) operating systems support, networking technologies, applications, and challenges: a comparative review. IEEE Commun. Surveys Tuts. **20**(3), 2062–2100 (2018)
4. V. Gazis, A survey of standards for machine-to-machine and the Internet of Things. IEEE Commun. Surveys Tuts. **19**(1), 482–511 (2017)
5. C. Yi, J. Cai, Two-stage spectrum sharing with combinatorial auction and stackelberg game in recall-based cognitive radio networks. IEEE Trans. Commun. **62**(11), 3740–3752 (2014)
6. G.-Y. Lin, H.-Y. Wei, A multi-period resource auction scheme for machine-to-machine communications, in *IEEE International Conference on Communication Systems (ICCS)* (IEEE, 2014), pp. 177–181
7. Y. Zhang, C. Lee, D. Niyato, P. Wang, Auction approaches for resource allocation in wireless systems: a survey. IEEE Commun. Surveys Tuts. **15**(3), 1020–1041 (2013)
8. N.V. Huynh, D.T. Hoang, D. Niyato, P. Wang, D.I. Kim, Optimal time scheduling for wireless-powered backscatter communication networks. IEEE Wireless Commun. Lett. **7**(5), 820–823 (2018)
9. H. Ju, R. Zhang, Throughput maximization in wireless powered communication networks. IEEE Trans. Wireless Commun. **13**(1), 418–428 (2014)
10. D.T. Hoang, D. Niyato, P. Wang, D.I. Kim, Z. Han, Ambient backscatter: a new approach to improve network performance for RF-powered cognitive radio networks. IEEE Trans. Commun. **65**(9), 3659–3674 (2017)
11. V. Liu, A. Parks, V. Talla, S. Gollakota, D. Wetherall, J.R. Smith, Ambient backscatter: wireless communication out of thin air, in *ACM SIGCOMM* (ACM, 2013), pp. 39–50
12. J. Gong, J.S. Thompson, S. Zhou, Z. Niu, Base station sleeping and resource allocation in renewable energy powered cellular networks. IEEE Trans. Commun. **62**(11), 3801–3813 (2014)

13. S. Zhong, J. Chen, Y. R. Yang, Sprite: a simple, cheat-proof, credit-based system for mobile ad-hoc networks,in *INFOCOM 2003. Twenty-Second Annual Joint Conference of the IEEE Computer and Communications* (IEEE, 2003), pp. 1987-1997
14. K. Chen, K. Nahrstedt, iPass: an incentive compatible auction scheme to enable packet forwarding service in MANET, in *24th International Conference on Distributed Computing Systems* (2004), pp. 534–542
15. G.L. Nemhauser, L.A. Wolsey, *Integer and Combinatorial Optimization* (Wiley, New York, NY, USA, 1988)
16. S. Boyd, J. Mattingley, Branch and bound methods, *EE364b course notes*, Stanford Univ., Stanford, CA, USA, Mar. (2007). http://www.stanford.edu/class/ee364b/lectures/bb_notes.pdf
17. S. Joshi, S. Boyd, Sensor selection via convex optimization. IEEE Trans. Signal Process. **57**(2), 451–462 (2009)
18. I. Sugathapala, M.F. Hanif, B. Lorenzo, S. Glisic, M. Juntti, L.-N. Tran, Topology adaptive sum rate maximization in the downlink of dynamic wireless networks. IEEE Trans. Commun. **66**(8), 3501–3516 (2018)
19. D. Lehmann, L.I. Ocallaghan, Y. Shoham, Truth revelation in approximately efficient combinatorial auctions. J. ACM **49**(5), 577–602 (2002)
20. S. Boyd, L. Vandenberghe, *Convex Optimization* (Cambridge University Press, Cambridge, U.K., 2004)
21. Z. Zheng, F. Wu, G. Chen, A strategy-proof combinatorial heterogeneous channel auction framework in noncooperative wireless networks. IEEE Trans. Mobile Comput. **14**(6), 1123–1137 (2015)
22. K. Yang, S. Martin, C. Xing, J. Wu, R. Fan, Energy-efficient power control for device-to-device communications. IEEE J. Sel. Areas Commun. **34**(12), 3208–3220 (2016)
23. Y. Jiao, P. Wang, S. Feng, D. Niyato, Profit maximization mechanism and data management for data analytics services. IEEE Internet Things J. **5**(3), 2001–2014 (2018)

Chapter 3
Contract-Based Time Assignment

In this chapter, we investigate the contract-based time assignment for backscatter-assisted RF-powered CR networks. Section 3.1 introduces the motivation of designing contract-based time assignment scheme and summarizes the contributions. Section 3.2 presents the system model, and Sect. 3.3 designs the optimal contract for the time resource assignment. Section 3.4 presents the simulation results, which are followed by the conclusions in Sect. 3.5.

3.1 Introduction

In backscatter-assisted RF-powered CR networks, both the HTT mode [1] and the backscatter mode [2] can be employed for the ST transmitting data to the SG [3, 4]. Since the backscatter communication is performed when the primary channel is busy, more time for backscatter communications results in less time for harvesting energy, which leads to a performance degradation in the HTT mode [4, 5]. Therefore, how much time resource should be assigned for transmitting data in the backscatter communication needs to be optimized.

It is notable that the ST does not need the participation of the SG when it harvests energy, but the ST needs the cooperation from the SG when the ST transmits data by the backscatter communications. Specifically, during the backscatter communication, the SG has to adapt its antenna from the sleep mode to the active mode and demodulate data from signals, which leads to the additional power consumption of the SG. To incentivize the SG to take part in the data transmission by the ST in the backscatter mode, pricing is an effective solution [5]. Specifically, the ST processes a certain payment to the SG, and the utilities of the ST and the SG can be improved from the integration of the backscatter communication. Nonetheless, this leads to a question that how much payment should be charged by the SG for the backscatter communication. Furthermore, the harvested power of the ST influences the network

© The Author(s), under exclusive license to Springer Nature Singapore Pte Ltd. 2021
X. Gao et al., *Resource Allocation in Backscatter-Assisted Communication Networks*,
https://doi.org/10.1007/978-981-16-5127-4_3

performance improvement from the integration of the backscatter communications. Although the ST can obtain the information of its harvested power, considering the fact that the SG is geographically separated from the ST, it is difficult for the SG to have this information. Therefore, the asymmetry of the harvested power of the ST should be considered when performing the backscatter time resource assignment and deriving the price. Contract theory is adopted as a powerful and effective tool to deal with the problem with asymmetric information [6, 7]. The above observations motivate us to design a contract for the time resource assignment in the backscatter-assisted RF-powered CR network.

In this chapter, firstly, we develop a contract model for the time resource assignment in the backscatter-assisted RF-powered CR network with asymmetric information, i.e., the ST has the information about the harvested power while the SG does not have. The SG designs a contract with multiple time-price items, and the ST accepts one contract item which can yield the largest utility to it. Secondly, we design an optimal contract, i.e., the optimal backscatter time resource vector and the optimal price vector, for the SG to maximize its profit. Note that our designed contract satisfies the incentive compatibility (IC) and the IR properties. Finally, numerical simulations are conducted to verify the effectiveness of our designed contract for the time resource assignment.

3.2   System Model

In this section, we first present the network model for the backscatter-assisted RF-powered CR network under consideration, and then develop the contract model for the time resource assignment in the network. Finally, we present the utility functions of the ST and the SG.

3.2.1   Network Model

We consider a backscatter-assisted RF-powered CR network, where there are a PT, an ST, and an SG, as shown in Fig. 3.1. The ST is within a coverage of the SG. The primary channel is divided into busy period and idle period, with normalized durations of α and $1 - \alpha$, respectively. When the primary channel is busy, the ST can (i) transmit data to the SG by backscattering the primary signals, or (ii) harvest energy. When the primary channel is idle, the ST can transmit data to the SG actively using the energy previously harvested. Therefore, the ST can transmit data to the SG in two modes. Specifically, transmitting data by backscattering the primary signals is referred to as the backscatter mode, and transmitting data with active RF signals using the energy previously harvested is referred to as the HTT mode [4]. Note that when there are multiple devices (STs perhaps SGs) in the backscatter-assisted RF-powered CR networks with asymmetric information of harvested powers, how the

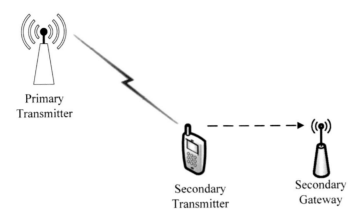

Fig. 3.1 Network model

resource is allocated among the devices needs to be investigated before employing our developed strategy in this chapter.

3.2.2 Contract Model

We present our contract model in Fig. 3.2. The SG designs a contract, including a series of time-price items, to maximize its profit. Accordingly, the SG can provide the service with different qualities, i.e., different backscatter time periods, by charging different prices. Among multiple contract items, the ST chooses the contract item which can maximize its utility [8, 9].

During the busy period, the ST can measure its harvested power, which is denoted as P_h. However, it is challenging for the SG to measure and know the value of P_h. Nevertheless, the statistical information of the value of P_h can be known by the SG according to the historical data. Therefore, according to the value of P_h, we classify the ST into different types. Note that the ST is more willing to harvest energy when P_h is larger. When P_h is no less than a threshold, which is denoted as $P_h^\dagger$, the ST is not willing to use the backscatter mode, as the backscatter communication cannot improve the amount of transmitted data. Therefore, we do not take the ST with $P_h \geq P_h^\dagger$ into account. Accordingly, when $P_h < P_h^\dagger$, the ST, whose harvested power is P_k^h, is referred to as a type-θ_k ST, where $\theta_k = 1/P_k^h$. We assume that the ST can be classified into K types when $P_h < P_h^\dagger$. Without loss of generality, P_k^h is sorted in a decreasing order, i.e., $P_1^h > \cdots > P_k^h > \cdots > P_K^h$. Accordingly, we have $\theta_1 < \cdots < \theta_k < \cdots < \theta_K$.

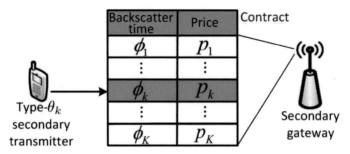

Fig. 3.2 Contract model

3.2.3 Utility Functions

We now present the utility functions of the ST and the SG. Let φ_k denote the time period of the type-θ_k ST transmitting data in the backscatter mode. Accordingly, the time period of the ST harvesting energy from the RF signals is $\alpha - \varphi_k$. From the analysis in [4, 5], the amount of the transmitted data per time unit by the type-θ_k ST can be expressed as

$$Z_k(\varphi_k) = R_b\varphi_k + \kappa B(1 - \alpha)\log_2\left(1 + \frac{\alpha - \varphi_k}{\theta_k P_0(1 - \alpha)}\right), \tag{3.1}$$

where R_b denotes the backscatter rate, B denotes the transmission bandwidth, P_0 is the ratio between the power of the noise and the channel gain from the ST to the SG, and κ denotes the efficiency of the transmission in the HTT mode. From (3.1), we observe that $Z_k(\varphi_k)$ is a concave function with respect to φ_k. Therefore, we can easily find the optimal backscatter time for the type-θ_k ST, which is denoted as $\varphi_k^\dagger$, to maximize $Z_k(\varphi_k)$. Considering the fact that the duration of the busy period is α, we have $\varphi_k^\dagger \leq \alpha$.

Let ϕ_k denote the designed backscatter time by the SG for the type-θ_k ST, i.e., the time when the antenna of the SG is active during the busy period and the SG is ready for receiving data in the backscatter mode. Note that ϕ_k may not be equal to φ_k. This is due to the fact that the ST does not need the cooperation from the SG and can be easily performed by itself when it harvests energy during the busy period. Therefore, with the predesigned ϕ_k, the maximum amount of the transmitted data by the type-θ_k ST is given by

$$R_k(\phi_k) = \begin{cases} Z_k(\phi_k) & \text{if } \phi_k \leq \varphi_k^\dagger, \\ Z_k(\varphi_k^\dagger) & \text{if } \phi_k > \varphi_k^\dagger. \end{cases} \tag{3.2}$$

Note that (3.2) indicates that the ST may choose a fraction of the time period to harvest energy although the antenna at the SG is active and the ST is allowed to transmit data in the backscatter mode during this period. This is due to the fact that using this fraction of time period to harvest energy can lead to more transmitted data per time unit. From (3.2), we can find that $R_k(\phi_k)$ is non-decreasing in ϕ_k. Accordingly, the valuation, i.e., the increased amount of the transmitted data due to the integration of backscatter communications, is given by

$$V_k(\phi_k) = R_k(\phi_k) - R_k(0). \tag{3.3}$$

Note that $V_k(\phi_k)$ is not always strictly increasing in ϕ_k. Let p_k denote the price which is corresponding to the backscatter time ϕ_k. The utility of the type-θ_k ST is defined as the difference between the increased amount of its transmitted data and its payment [5], which is given by

$$u_{\text{st},k}(\phi_k, p_k) = V_k(\phi_k) - p_k. \tag{3.4}$$

Accordingly, the utility of the SG, which is designed for the type-θ_k ST, is given by

$$u_{\text{sg},k}(\phi_k, p_k) = p_k - c\phi_k, \tag{3.5}$$

where $c > 0$ is the cost per unit time when the SG receives the data in the backscatter mode.

3.3 Optimal Contract Design

In this section, we design the optimal contract for the time resource assignment in the backscatter-assisted RF-powered CR network. We first formulate the optimal contract problem, and then derive the optimal pricing scheme and the optimal backscatter time sequentially.

3.3.1 Contract Formulation

A feasible contract should guarantee the IC and the IR properties for all types [8]. The IC property means that the ST choosing the contract item which is not designed for its type cannot increase its utility [8]. Therefore, the IC property can guarantee that the ST will choose the contract item which is designed for it, instead of the other contract items. To guarantee the IC property of our designed contract, the utility of the type-θ_k ST with a choice of $\{\phi_k, p_k\}$ should be no less than that with a choice of $\{\phi_{k'}, p_{k'}\}$, $\forall k' \neq k$. Accordingly, it can be expressed as

$$u_{\text{st},k}(\phi_k, p_k) \geq u_{\text{st},k}(\phi_{k'}, p_{k'}), k' \neq k, 1 \leq k \leq K. \tag{3.6}$$

The IR property means that the utility of the type-θ_k ST choosing the contract item which is designed for its type is non-negative [8]. The ST will not accept the contract if the utility of choosing the contract item is negative. Accordingly, it can be expressed as

$$u_{\text{st},k}(\phi_k, p_k) \geq 0, 1 \leq k \leq K. \tag{3.7}$$

Taking the IC and the IR properties into account, we formulate the optimal contract problem for the time resource assignment and pricing in the network as

$$\mathcal{P}_{3.1} : \quad \max_{\phi, p} \sum_{k=1}^{K} f_k u_{\text{sg},k}(\phi_k, p_k) \tag{3.8a}$$

$$s.t. \ (3.6), (3.7), \text{ and}$$

$$0 \leq \phi_k \leq \alpha, 1 \leq k \leq K. \tag{3.8b}$$

where $\phi = \{\phi_k\}_{k=1}^{K}$, $p = \{p_k\}_{k=1}^{K}$, and f_k is the possibility of the ST with type-θ_k. Let $\{\phi^*, p^*\}$ denote the optimal solution to $\mathcal{P}_{3.1}$, where $\phi^* = \{\phi_k^*\}_{k=1}^{K}$ and $p^* = \{p_k^*\}_{k=1}^{K}$. Due to the non-convexity of the constraint in (3.6), $\mathcal{P}_{3.1}$ is a nonconvex optimization problem. Therefore, we sequentially derive the optimal price, i.e. p^*, and the optimal backscatter time, i.e., ϕ^*, in the following.

3.3.2   Optimal Pricing

Before deriving the optimal pricing scheme, we first introduce two lemmas. Firstly, we transform the constraint in (3.8b) into a more strict one, and introduce the following lemma.

Lemma 3.1 *For the optimal solution to $\mathcal{P}_{3.1}$, we have $\phi_k^* \leq \varphi_k^{\dagger}, 1 \leq k \leq K$.*

Proof We prove Lemma 3.1 by contradiction. Assume that there exists $\phi_{\hat{k}}^* > \varphi_{\hat{k}}^{\dagger}, \forall \hat{k} \in \hat{\mathcal{K}}$, in the optimal solution $\{\phi^*, p^*\}$, where $\hat{\mathcal{K}}$ is a non-empty subset of $\{1, 2, \dots, K\}$. Consider a backscatter time vector $\bar{\phi} = \{\bar{\phi}_k\}_{k=1}^{K}$, where $\bar{\phi}_k = \min\{\phi_k^*, \varphi_k^{\dagger}\}$. From (3.4), we find that $u_{\text{st},k}(\phi_k, p_k)$ is non-decreasing in ϕ_k, and $u_{\text{st},k}(\phi_k, p_k)$ remains unchanged when $\phi_k \geq \varphi_k^{\dagger}$. Therefore, we have

$$u_{\text{st},k}(\bar{\phi}_k, p_k^*) = u_{\text{st},k}(\phi_k^*, p_k^*), 1 \leq k \leq K \tag{3.9}$$

and

$$u_{\text{st},k}(\bar{\phi}_{k'}, p_{k'}^*) \leq u_{\text{st},k}(\phi_{k'}^*, p_{k'}^*), k' \neq k, 1 \leq k \leq K. \tag{3.10}$$

From (3.6), (3.7), (3.9), and (3.10), we can obtain

$$u_{\text{st},k}(\bar{\phi}_k, p_k^*) \geq u_{\text{st},k}(\bar{\phi}_{k'}, p_{k'}^*), k' \neq k, 1 \leq k \leq K \tag{3.11}$$

and

$$u_{\text{st},k}(\bar{\phi}_k, p_k^*) \geq 0, 1 \leq k \leq K. \tag{3.12}$$

Hence, $\{\bar{\phi}, p^*\}$ also guarantees the IC and the IR properties, and it is also a feasible solution to $\mathcal{P}_{3.1}$. Since $\phi_{\hat{k}}^* > \varphi_{\hat{k}}^\dagger, \forall \hat{k} \in \hat{\mathcal{K}}$, we have

$$u_{\text{sg},\hat{k}}(\bar{\phi}_{\hat{k}}, p_{\hat{k}}^*) > u_{\text{sg},\hat{k}}(\phi_{\hat{k}}^*, p_{\hat{k}}^*), \forall \hat{k} \in \hat{\mathcal{K}}. \tag{3.13}$$

Therefore, it follows that

$$\sum_{k=1}^{K} f_k u_{\text{sg},k}(\bar{\phi}_k, p_k^*) > \sum_{k=1}^{K} f_k u_{\text{sg},k}(\phi_k^*, p_k^*) \tag{3.14}$$

Note that the inequality in (3.14) contradicts the fact that $\{\phi_k^*, p_k^*\}$ is an optimal solution to $\mathcal{P}_{3.1}$, which completes the proof of this lemma. ∎

From Lemma 3.1, the constraint in (3.8b) can be transformed to

$$0 \leq \phi_k \leq \varphi_k^\dagger, 1 \leq k \leq K. \tag{3.15}$$

Accordingly, $\mathcal{P}_{3.1}$ can be equivalently transformed to

$$\mathcal{P}_{3.2}: \quad \max_{\phi, p} \sum_{k=1}^{K} f_k u_{\text{sg},k}(\phi_k, p_k)$$

$$\text{s.t. (3.6), (3.7), and (3.15).}$$

After introducing the constraint in (3.15), we present the increasing preference (IP) property in Lemma 3.2. Specifically, with the same increase of the backscatter time, a higher type ST can bring more valuation than a lower type ST.

Lemma 3.2 *For any type $\theta_{k'} < \theta_k$ and any backscatter time $\phi' < \phi \leq \varphi_k^\dagger$, we have $V_k(\phi) - V_k(\phi') > V_{k'}(\phi) - V_{k'}(\phi')$.*

Proof We prove Lemma 3.2 in the following three cases: (i) $\phi' < \phi \leq \varphi_{k'}^\dagger$, (ii) $\phi' < \varphi_{k'}^\dagger < \phi$, and (iii) $\varphi_{k'}^\dagger \leq \phi' < \phi$. For notational brevity, let $\varepsilon = \frac{\alpha - \phi}{(1-\alpha)P_0}$ and $\varepsilon' = \frac{\alpha - \phi'}{(1-\alpha)P_0}$. Accordingly, we have $\varepsilon < \varepsilon'$, and

$$V_k(\phi) - V_k(\phi') = R_b(\phi - \phi') + \kappa B(1 - \alpha) \log_2\left(1 + \frac{\varepsilon - \varepsilon'}{\theta_k + \varepsilon'}\right). \tag{3.16}$$

Case (i): When $\phi' < \phi \leq \varphi_{k'}^{\dagger}$, we have

$$V_{k'}(\phi) - V_{k'}(\phi') = R_b(\phi - \phi') + \kappa B(1 - \alpha) \log_2 \left(1 + \frac{\varepsilon - \varepsilon'}{\theta_{k'} + \varepsilon'}\right). \qquad (3.17)$$

Since $\theta_{k'} < \theta_k$, from (3.16) and (3.17), it follows that $V_k(\phi) - V_k(\phi') > V_{k'}(\phi) - V_{k'}(\phi')$.

Case (ii): When $\phi' < \varphi_{k'}^{\dagger} < \phi$, from the result in Case (i), we have

$$V_k(\varphi_{k'}^{\dagger}) - V_k(\phi') > V_{k'}(\varphi_{k'}^{\dagger}) - V_{k'}(\phi'). \qquad (3.18)$$

In addition, when $\phi' < \varphi_{k'}^{\dagger} < \phi \leq \varphi_k^{\dagger}$, from (3.3), we have $V_k(\phi) > V_k(\varphi_{k'}^{\dagger})$ and $V_{k'}(\phi) = V_{k'}(\varphi_{k'}^{\dagger})$. As such, recalling (3.18), it follows that $V_k(\phi) - V_k(\phi') > V_{k'}(\phi) - V_{k'}(\phi')$.

Case (iii): When $\varphi_{k'}^{\dagger} \leq \phi' < \phi$, we have $V_{k'}(\phi) = V_{k'}(\phi') = V_{k'}(\varphi_{k'}^{\dagger})$, from which we can obtain $V_{k'}(\phi) - V_{k'}(\phi') = 0$. In addition, from (3.3), we have $V_k(\phi) - V_k(\phi') > 0$ when $\phi' < \phi \leq \varphi_k^{\dagger}$. Therefore, it follows that $V_k(\phi) - V_k(\phi') > V_{k'}(\phi) - V_{k'}(\phi')$.

We now complete the proof of Lemma 3.2.   ∎

With the IP property from Lemma 3.2, according to [8], the optimal pricing scheme can be obtained by the following proposition.

Proposition 3.1 *When the valuation of the ST satisfies the IP property, to derive a feasible contract guaranteeing the IC and the IR properties, we have:*

1. *A necessary condition of a feasible contract is $\phi_1 \leq \phi_2 \leq \cdots \leq \phi_K$.*
2. *With the necessary condition $\phi_1 \leq \phi_2 \leq \cdots \leq \phi_K$, the optimal pricing scheme is*

$$p_k^* = \begin{cases} V_k(\phi_k) & \text{if } k = 1, \\ p_{k-1}^* + V_k(\phi_k) - V_k(\phi_{k-1}) & \text{if } 2 \leq k \leq K. \end{cases} \qquad (3.19)$$

Based on (3.19) in Proposition 3.1, when the backscatter time vector is determined, the optimal pricing scheme can be given by

$$p_k^* = \begin{cases} V_k(\phi_k) & \text{if } k = 1, \\ V_1(\phi_1) + \sum_{i=2}^{k} (V_i(\phi_i) - V_i(\phi_{i-1})) & \text{if } 2 \leq k \leq K. \end{cases} \qquad (3.20)$$

3.3.3 Optimal Backscatter Time

In the following, we substitute the optimal pricing result (3.20) into $\mathcal{P}_{3.2}$, and derive the optimal backscatter time vector by the following problem:

$$\mathcal{P}_{3.3} : \max_{\phi} V_1(\phi_1) + \sum_{k=2}^{K} f_k \sum_{i=2}^{k} (V_i(\phi_i) - V_i(\phi_{i-1})) - c \sum_{k=1}^{K} f_k \phi_k \qquad (3.21a)$$

$$s.t. \ (3.15), \ \text{and}$$

$$\phi_1 \leq \phi_2 \leq \cdots \leq \phi_K. \qquad (3.21b)$$

Note that the objective function in $\mathcal{P}_{3.3}$ is not concave. To derive a solution to $\mathcal{P}_{3.3}$, we rearrange the items in (3.21a), and rewrite (3.21a) as

$$\sum_{k=1}^{K} J_k(\phi_k) \qquad (3.22)$$

where

$$J_k(\phi_k) = \begin{cases} \sum_{i=k}^{K} f_i V_k(\phi_k) - \sum_{i=k+1}^{K} f_i V_{k+1}(\phi_k) - cf_k\phi_k, & \text{if } 1 \leq k \leq K-1, \\ f_k V_k(\phi_k) - cf_k\phi_k, & \text{if } k = K. \end{cases}$$
$$(3.23)$$

From (3.22) and (3.23), the objective function in $\mathcal{P}_{3.3}$ is rewritten as the sum of K functions, and each function $J_k(\phi_k)$ is only related to ϕ_k. Therefore, maximizing the objective function in $\mathcal{P}_{3.3}$ is equivalent to maximizing each $J_k(\phi_k)$. Since $J_k(\phi_k)$ is a function with only one variable, the optimal solution to ϕ_k is obtained at the boundary point or the stationary points satisfying the constraint in (3.15), which can be derived by letting the first-order derivative of $J_k(\phi_k)$, i.e., $\frac{dJ_k(\phi_k)}{d\phi_k}$, equal to zero. Among these points, we choose the point with maximum value of $J_k(\phi_k)$ as the solution. Note that $\frac{dJ_k(\phi_k)}{d\phi_k} = 0$ is a quadratic equation (when $1 \leq k \leq K-1$) or a linear equation (when $k = K$) with one variable, which is very easy to solve. If the obtained solution satisfies the constraint in (3.21b), the obtained solution is the optimal backscatter time. If the obtained solution violates the constraint in (3.21b), the dynamic algorithm in [8] will be employed to obtain a backscatter time solution which satisfies the constraint in (3.21b). After obtaining the backscatter time result, the optimal price can be obtained based on (3.20).

We now present how to implement the developed scheme in the backscatter-assisted RF-powered CR network. After the SG obtains the statistical information about the harvested power of the ST, which can be obtained in advance from the historical data, the SG designs a contract by solving $\mathcal{P}_{3.3}$, and broadcasts the designed contract to the ST. Next, the ST sends its selected contract information to the SG, i.e., signs the contract with the SG. Finally, the SG demodulates the transmitted data from the ST and charges the price to the ST according to the signed contract. From the above process, we find that in the contract design, the SG derives the solution to $\mathcal{P}_{3.3}$, which is easy to obtain, broadcasts the designed contract, and charges the price. The ST sends the several-bit selected contract information to the SG. Overall, the complexity of our designed scheme is very low, which is suitable for the backscatter-assisted RF-powered CR network.

3.4   Simulation Results

We evaluate the performance of our developed scheme by using MATLAB. Specifically, we consider a backscatter-assisted RF-powered CR network, where the normalized busy period and the normalized idle period are 0.8 and 0.2, respectively, the bandwidth of the channel is 100 kHz, the efficiency of the transmission in the HTT mode is 0.6, and the backscatter rate is 30 kbps [5]. The harvested power of the type-θ_1 ST (i.e., P_1^h) is 5 μW, and the difference of the harvested power of the ST between two adjacent types (i.e., $P_k^h - P_{k+1}^h$) is 0.1 μW. The cost per time unit when the SG receives the data in the backscatter mode is 10^3. In addition, similar to [9, 10], we consider that the probability of the ST with different types follows the uniform distribution.

We verify the IC and the IR properties of our developed contract in Fig. 3.3. In this figure, we consider an ST with 20 types, and plot the utility of the ST when the ST with type-θ_{10}, type-θ_{12}, or type-θ_{14}, chooses different contract items. From this figure, we observe that the utility of the ST choosing the contract item corresponding to its own type is no less than that of the ST choosing the other contract items, indicating that our designed contract guarantees the IC property. We also find that the utility of the ST choosing the corresponding contract item is no less than zero, which demonstrates that our designed contract guarantees the IR property.

We next compare our developed contract-based scheme with the linear pricing scheme. In the linear pricing scheme, the SG announces a price per time unit for transmitting data in the backscatter mode, and the ST optimizes its utility. Note that

Fig. 3.3 Verify the IC and the IR properties

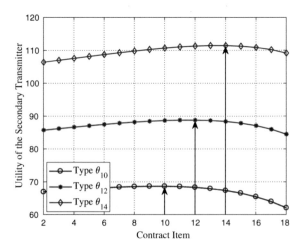

Fig. 3.4 Utility of the SG
versus the unit price in the
linear pricing scheme

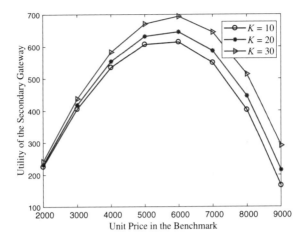

different unit prices in the linear pricing scheme will lead to different utilities of the SG. We plot the utility of the SG with different unit prices in the linear pricing in Fig. 3.4. From this figure, we find that the utility of the SG is maximized when the unit price is around 6×10^3. Therefore, we set the unit price in the linear pricing scheme as 6×10^3 when we compare our developed contract-based scheme with the linear pricing scheme. Figure 3.5 shows the utility of the SG in the two schemes. From this figure, we observe that the utility of the SG in our developed scheme is higher than that in the linear pricing scheme. Figure 3.6 presents the average increased amount of the ST's transmitted data due to the integration of the backscatter communications in the two schemes. This figure shows that the average increased amount of the ST's transmitted data in our developed contract-based scheme is much higher than that in the linear pricing scheme. The results in Figs. 3.5 and 3.6 demonstrate that

Fig. 3.5 The comparison of
the utility of the SG between
our developed contract-based
scheme and the linear pricing
scheme

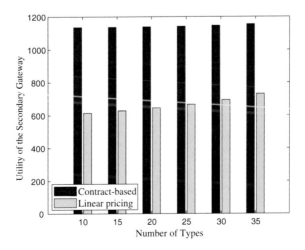

Fig. 3.6 The comparison of the average increased amount of the ST's transmitted data between our developed contract-based scheme and the linear pricing scheme

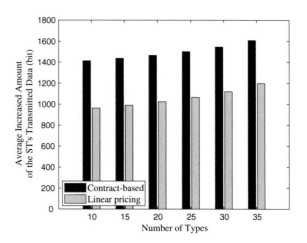

our developed contract-based scheme outperforms the linear pricing scheme in the scenario with asymmetric information.

3.5   Conclusions

In this chapter, we have considered the asymmetric information of the harvested power of the ST and designed an incentive time resource assignment and pricing scheme for the backscatter-assisted RF-powered CR network based on contract theory. According to the value of the harvested power of the ST, the ST is classified into several types. The SG considers the statistical property of the types and announces a contract to assign the backscatter time resource for the ST. Then, the ST accepts one contract item to maximize its utility. We have derived the optimal contract for the SG, and shown that the designed contract satisfies the IC and the IR properties, which has been verified by the numerical results. We have also demonstrated that the proposed time resource assignment and pricing scheme outperforms the linear pricing strategy, which has provided useful guidance for the design of in backscatter-assisted RF-powered CR networks.

References

1. H. Ju, R. Zhang, Throughput maximization in wireless powered communication networks. IEEE Trans. Wireless Commun. **13**(1), 418–428 (2014)
2. V. Liu, A. Parks, V. Talla, S. Gollakota, D. Wetherall, J.R. Smith, Ambient backscatter: wireless communication out of thin air, in *ACM SIGCOMM* (ACM, 2013), pp. 39–50
3. D.T. Hoang, D. Niyato, P. Wang, D.I. Kim, Zhu Han, The tradeoff analysis in RF-powered backscatter cognitive radio networks. in *IEEE Globecom* (IEEE, 2016), pp. 1–6

4. D.T. Hoang, D. Niyato, P. Wang, D.I. Kim, Z. Han, Ambient backscatter: a new approach to improve network performance for RF-powered cognitive radio networks. IEEE Trans. Commun. **65**(9), 3659–3674 (2017)
5. D.T. Hoang, D. Niyato, P. Wang, D.I. Kim, L.B. Le, Overlay RF-powered backscatter cognitive radio networks: a game theoretic approach, in *IEEE ICC* (IEEE, 2017), pp. 1–6
6. Y. Zhang, M. Pan, L. Song, Z. Dawy, Zhu Han, A survey of contract theory-based incentive mechanism design in wireless networks. IEEE Wireless Commun. **24**(3), 80–85 (2017)
7. Z. Hu, Z. Zheng, L. Song, T. Wang, X. Li, UAV offloading: spectrum trading contract design for UAV-assisted cellular networks. IEEE Trans. Wireless Commun. **17**(9), 6093–6107 (2018)
8. L. Gao, X. Wang, Y. Xu, Q. Zhang, Spectrum trading in cognitive radio networks: a contract-theoretic modeling approach. IEEE J. Sel. Areas Commun. **29**(4), 843–855 (2011)
9. Y. Zhang, L. Song, W. Saad, Z. Dawy, Z. Han, Contract-based incentive mechanisms for device-to-device communications in cellular networks. IEEE J. Sel. Areas Commun. **33**(10), 2144–2155 (2015)
10. M. Zeng, Y. Li, K. Zhang, M. Waqas, D. Jin, Incentive mechanism design for computation offloading in heterogeneous fog computing: a contract-based approach, in *IEEE ICC* (IEEE, 2018), pp. 1–6

Chapter 4
Evolutionary Game-Based Access Point and Service Selection

In this chapter, we investigate the evolutionary game-based AP and service selection for backscatter-assisted RF-powered CR networks. Section 4.1 introduces the motivation of developing evolutionary game-based AP and service selection and summarizes the contributions. Section 4.2 presents the system model. Section 4.3 formulates the AP and service selection as an evolutionary game, and analyzes the evolutionary equilibrium. Section 4.4 derives the stability region of the delayed replicator dynamics in a special case. Section 4.5 develops an algorithm to implement the AP and service selection in the network. Section 4.6 presents the numerical results, and Sect. 4.7 concludes this chapter.

4.1 Introduction

In backscatter-assisted RF-powered CR networks, the backscatter mode [1] and the HTT mode [2] can be utilized for the STs transmitting data [3, 4]. In addition, many STs may need to choose one of available APs in the networks to transmit data. As such, the time resource should be assigned to the STs carefully. The APs are responsible for scheduling and assigning the time resource to the STs to transmit data, and pricing is an effective method to stimulate the APs to allocate the time resource for the STs. To satisfy different QoS requirements, the APs usually provide different services for the STs with different prices. It is notable that the STs choosing different APs and services may achieve different utilities, and the STs with lower utilities have the incentives to adapt their AP and service selections [5–7]. Therefore, the AP and service selection strategy for the STs changes over time.

Game theory can model the devices making independent and rational decisions, and can develop low-complexity distributed algorithms to describe the relationships

© The Author(s), under exclusive license to Springer Nature Singapore Pte Ltd. 2021
X. Gao et al., *Resource Allocation in Backscatter-Assisted Communication Networks*,
https://doi.org/10.1007/978-981-16-5127-4_4

among different entities. Hence, game theory is an attractive tool for developing and analyzing distributed, flexible, and autonomous networks [5]. Among various game-theoretic approaches, evolutionary game, which can model the strategy adaptation of players, is a suitable tool to deal with the problem with dynamic strategies [8]. In addition, the players in the evolutionary game are bounded rational, and they adapt their strategies gradually to reach the evolutionary equilibrium [6]. Furthermore, the algorithm to implement the strategy adaptation based on the evolutionary game has a low complexity [6, 8], which is suitable for the power-constrained systems, e.g., backscatter-assisted RF-powered CR networks.

Motivated by the above observations, in this chapter, we investigate the dynamic AP and service selection in a backscatter-assisted RF-powered CR network based on evolutionary game. The main contributions of this chapter are summarized as follows.

1. We formulate the AP and service selection in a backscatter-assisted RF-powered CR network as an evolutionary game. In the evolutionary game, the STs act as players and form a population, and they adjust their selections of the APs and services based on their utilities.
2. We model the AP and service adaptation of the STs as replicator dynamics in the evolutionary game. An evolutionary equilibrium is eventually reached based on the replicator dynamics. We analytically prove the existence and uniqueness of the evolutionary equilibrium. We also prove that the obtained evolutionary equilibrium is stable.
3. We take the delay of the information used by the STs into account and model the AP and service adaptation by the delayed replicator dynamics. We show that with a small delay, an evolutionary equilibrium can be also reached. Also, we analytically derive the stability region of the delayed replicator dynamics in a special case.
4. We further develop an algorithm to implement the AP and service selection based on evolutionary game. The developed algorithm is computationally efficient and also scalable, which is suitable for the backscatter-assisted RF-powered CR network. By using the algorithm, the experimental results clearly show the consistency with the analytical results, validating the developed game models.

4.2   System Model

In this section, we present the system model of the backscatter-assisted RF-powered CR network under consideration. We then define the utilities of the STs choosing different APs and services.

4.2.1 Network Model

We consider a backscatter-assisted RF-powered CR network, as shown in Fig. 4.1. In the network, the PT transmits signals on a periodic basis, where the normalized busy and idle periods are denoted as α and $1 - \alpha$, respectively. There exist M APs in the network, and we denote the set of the APs as $\mathcal{M} = \{1, 2, \ldots, M\}$. There are a group of STs which are within a coverage of the APs, and the number of the STs is N.

In the backscatter-assisted RF-powered CR network, the STs can transmit data to the APs in the HTT mode and/or the backscatter mode. Therefore, each ST can transmit data in one of the three modes: (i) HTT only mode, (ii) hybrid mode (i.e., the combination of the HTT mode and the backscatter mode), and (iii) backscatter only mode. Each AP can provide the STs different qualities of services, each of which is implemented in one of the three modes, with different prices. Without loss of generality and for an ease of presentation, we assume that each AP can provide 3 services for the STs transmitting data in the above three modes, respectively. Specifically, in Service 1, the STs can transmit data only in the HTT mode. In Service 2, the STs can transmit data in the hybrid mode, and in Service 3, the STs can transmit data only in the backscatter mode. Therefore, Service 1 and Service 3 are the special cases of Service 2. Note that although we consider only 3 services in our work, our work can be easily extended to the case where each AP can provide more than 3 services.

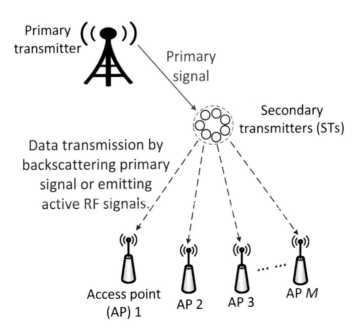

Fig. 4.1 System model of the backscatter-assisted RF-powered CR network

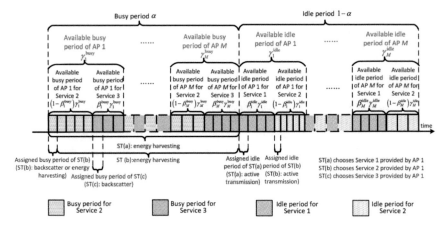

Fig. 4.2 Time resource assignment during one time unit

Figure 4.2 presents the time resource assignment during one time unit. In the secondary network, the TDMA is adopted [2]. After the STs choose APs and services, the APs broadcast the assigned time period to the STs. Therefore, at a time, only one AP is available, and only one ST is allowed to transmit data. Hence, there is no interference among the STs. In addition, each AP partitions its available idle time resource into two parts for Service 1 and Service 2, respectively, and partitions its available busy time resource into two parts for Service 2 and Service 3, respectively. Let γ_m^{idle} and γ_m^{busy}, $m \in \mathcal{M}$, denote the normalized available idle and busy periods of AP m, respectively. Let β_m^{idle} denote the fraction of the idle period for Service 1 at AP m. Accordingly, the normalized idle periods for Service 1 and Service 2 at AP m are $\beta_m^{idle}\gamma_m^{idle}$ and $(1 - \beta_m^{idle})\gamma_m^{idle}$, respectively. Let β_m^{busy} denote the fraction of the busy period for Service 3 at AP m. Accordingly, the normalized busy periods for Service 2 and Service 3 at AP m are $(1 - \beta_m^{busy})\gamma_m^{busy}$ and $\beta_m^{busy}\gamma_m^{busy}$, respectively. During one time unit, each ST transmits data to one selected AP by using one of the three services. Let $N_{m,i}$ denote the number of STs choosing Service i provided by AP m. Accordingly, the proportion of the STs choosing Service i provided by AP m can be expressed as $x_{m,i} = N_{m,i}/N$.

For a given service provided by a given AP, the time period is equally assigned by the AP to the STs choosing the same AP and service [6]. Accordingly, the assigned idle period for each ST choosing Service 1 provided by AP m is

$$\tau_{m,1}^{idle} = \frac{\beta_m^{idle}\gamma_m^{idle}}{x_{m,1}N}. \tag{4.1}$$

Similarly, the assigned idle period and the assigned busy period for each ST choosing Service 2 provided by AP m are

$$\tau_{m,2}^{\text{idle}} = \frac{(1 - \beta_m^{\text{idle}})\gamma_m^{\text{idle}}}{x_{m,2} N} \tag{4.2}$$

and

$$\tau_{m,2}^{\text{busy}} = \frac{(1 - \beta_m^{\text{busy}})\gamma_m^{\text{busy}}}{x_{m,2} N}, \tag{4.3}$$

respectively, and the assigned busy period for each ST choosing Service 3 provided by AP m is

$$\tau_{m,3}^{\text{busy}} = \frac{\beta_m^{\text{busy}}\gamma_m^{\text{busy}}}{x_{m,3} N}. \tag{4.4}$$

In Fig. 4.2, we take ST (a), ST (b), and ST (c), which choose Service 1 provided by AP 1, Service 2 provided by AP 1, and Service 3 provided by AP 1, respectively, as examples to illustrate how the STs transmit data to the APs. When the STs (e.g., ST (a)) choose Service 1, each of the STs harvests energy during the whole busy period, and transmits data to its selected AP by active transmission during its assigned idle period. When the STs (e.g., ST (b)) choose Service 2, each of the STs can choose to backscatter or harvest energy during its assigned busy period. Note that the STs choosing Service 2 may not utilize all the assigned busy period for the backscatter communications, since an increase of the backscatter time may reduce the total amount of the transmitted data. When the STs choose Service 2, each of the STs harvests energy from the primary signals during its unassigned busy period (i.e., the busy period except its assigned busy period), and transmits data to its selected AP by active transmission during its assigned idle period. When the STs (e.g., ST (c)) choose Service 3, each of the STs transmits data by using backscatter communications during its assigned busy period. Considering that the APs usually have constant available energy sources, we assume that the APs have enough energy for receiving data.

4.2.2 Utility Functions

We now present the utility functions of the STs choosing different APs and services. We assume that the STs are statistically identical in location and propagation environment [9]. Therefore, the STs choosing the same AP and the same service have the same statistical channel state information, and the STs choosing the same AP and the same service have the same utilities. Specifically, when the STs choose Service 1, only the HTT mode can be employed by the STs to transmit data. Therefore, the normalized energy harvesting period of each ST choosing Service 1 is α. Accordingly, the harvested energy of each ST choosing Service 1 is

$$E_1^{h} = P_{h}\alpha, \tag{4.5}$$

where P_{h} is the harvested power of one ST. Therefore, the amount of the transmitted data of each ST choosing Service 1 provided by AP m during one time unit is given by

$$Z_{m,1} = \tau_{m,1}^{\text{idle}} \xi B \log_2 \left(1 + \frac{g_m E_1^{h}}{\sigma^2 \tau_{m,1}^{\text{idle}}} \right), \tag{4.6}$$

where B is the bandwidth of the channel, ξ is the efficiency of the active transmission, g_m is the channel gain between the STs and AP m, and σ^2 is the power of the noise. Note that g_m can be viewed as the equivalent channel gain between the STs and AP m when the STs only have the identically statistical channel state information and they transmit data with a certain reliability [10, 11]. Substituting (4.5) into (4.6), $Z_{m,1}$ can be rewritten as

$$Z_{m,1} = \tau_{m,1}^{\text{idle}} \xi B \log_2 \left(1 + \frac{\varsigma_m \alpha}{\tau_{m,1}^{\text{idle}}} \right), \tag{4.7}$$

where $\varsigma_m = g_m P_{h}/\sigma^2$.

When the STs choose Service 2, both the backscatter mode and the HTT mode can be employed by the STs to transmit data to the APs. In the backscatter mode, the transmission rate of the STs to AP m is denoted as R_m^{b}. Accordingly, the amount of the transmitted data of each ST choosing Service 2 provided by AP m in the backscatter mode during one time unit is

$$Z_{m,2}^{b} = R_m^{b} \tau_{m,2}^{b}, \tag{4.8}$$

where $\tau_{m,2}^{b}$ denotes the actual backscattering time period of each ST choosing Service 2 provided by AP m.

During the time period of one ST backscattering primary signals, the ST cannot harvest the energy which is used for the data transmission in the HTT mode [4]. Therefore, the normalized energy harvesting period of each ST choosing Service 2 provided by AP m is $\alpha - \tau_{m,2}^{b}$. Accordingly, the harvested energy of each ST choosing Service 2 provided by AP m is

$$E_{m,2}^{h} = P_{h}(\alpha - \tau_{m,2}^{b}). \tag{4.9}$$

Therefore, the amount of the transmitted data of each ST choosing Service 2 provided by AP m during one time unit is given by

$$Z_{m,2}^{t} = \tau_{m,2}^{\text{idle}} \xi B \log_2 \left(1 + \frac{g_m E_{m,2}^{h}}{\sigma^2 \tau_{m,2}^{\text{idle}}} \right). \tag{4.10}$$

Substituting (4.9) into (4.10), $Z^t_{m,2}$ can be rewritten as

$$Z^t_{m,2} = \tau^{idle}_{m,2} \xi B \log_2 \left(1 + \frac{\varsigma_m (\alpha - \tau^b_{m,2})}{\tau^{idle}_{m,2}} \right). \tag{4.11}$$

From (4.8) to (4.11), we find that an increase of the backscatter time will reduce the time period of harvesting energy, which results in the reduction of the amount of the transmitted data in the HTT mode. Therefore, the STs choosing Service 2 may not utilize all the assigned busy period for backscatter communication. Accordingly, the maximum amount of the transmitted data of each ST choosing Service 2 provided by AP m during one time unit can be expressed as

$$Z_{m,2} = \max_{\tau^b_{m,2}} Z^b_{m,2} + Z^t_{m,2},$$

$$s.t. \ 0 \leq \tau^b_{m,2} \leq \tau^{busy}_{m,2}. \tag{4.12}$$

Note that the optimization problem in (4.12) is a convex optimization problem, and a similar problem has been solved in [4]. Specifically, performing the first-order derivative of $(Z^b_{m,2} + Z^t_{m,2})$ with respect to $\tau^b_{m,2}$, we have

$$\frac{d(Z^b_{m,2} + Z^t_{m,2})}{d\tau^b_{m,2}} = R^b_m - \frac{\tau^{idle}_{m,2} \xi B \varsigma_m}{(\ln 2)(\tau^{idle}_{m,2} + \varsigma_m (\alpha - \tau^b_{m,2}))}. \tag{4.13}$$

We denote $\tau^{b,\dagger}_{m,2}$ as the root of (4.13). Accordingly, let $\frac{d(Z^b_{m,2}+Z^t_{m,2})}{d\tau^b_{m,2}} = 0$, and we can obtain

$$\tau^{b,\dagger}_{m,2} = \alpha - \tau^{idle}_{m,2} \left(\frac{\xi B}{R^b_m (\ln 2)} - \frac{1}{\varsigma_m} \right). \tag{4.14}$$

Therefore, the optimal objective value of the optimization problem in (4.12) can be obtained when

$$\tau^b_{m,2} = \begin{cases} 0, & \text{if } \tau^{b,\dagger}_{m,2} \leq 0, \\ \tau^{b,\dagger}_{m,2}, & \text{if } 0 < \tau^{b,\dagger}_{m,2} < \tau^{busy}_{m,2}, \\ \tau^{busy}_{m,2}, & \text{if } \tau^{b,\dagger}_{m,2} \geq \tau^{busy}_{m,2}. \end{cases} \tag{4.15}$$

Accordingly, $Z_{m,2}$ can be rewritten as

$$Z_{m,2} = \begin{cases} \tau_{m,2}^{idle}\xi B \log_2\left(1 + \frac{\varsigma_m\alpha}{\tau_{m,2}^{idle}}\right), & \text{if } \tau_{m,2}^{b,\dagger} \leq 0, \\ R_m^b\tau_{m,2}^{b,\dagger} + \tau_{m,2}^{idle}\xi B \log_2\left(1 + \frac{\varsigma_m(\alpha-\tau_{m,2}^{b,\dagger})}{\tau_{m,2}^{idle}}\right), & \text{if } 0 < \tau_{m,2}^{b,\dagger} < \tau_{m,2}^{busy}, \\ R_m^b\tau_{m,2}^{busy} + \tau_{m,2}^{idle}\xi B \log_2\left(1 + \frac{\varsigma_m(\alpha-\tau_{m,2}^{busy})}{\tau_{m,2}^{idle}}\right), & \text{if } \tau_{m,2}^{b,\dagger} \geq \tau_{m,2}^{busy}. \end{cases} \quad (4.16)$$

When the STs choose Service 3, the STs can transmit data only in the backscatter mode. Therefore, the amount of the transmitted data of each ST choosing Service 3 provided by AP m during one time unit is

$$Z_{m,3} = R_m^b\tau_{m,3}^{busy}. \quad (4.17)$$

To guarantee different QoS requirements of the STs, different services provided by different APs are marked with different prices. Let $p_{m,i}$ denote the price charged by AP m which provides Service i for one ST. Accordingly, the utility of each ST choosing Service i provided by AP m, defined as the ratio of the amount of the transmitted data during one time unit and the price [9], can be expressed as

$$u_{m,i} = \frac{Z_{m,i}}{p_{m,i}}. \quad (4.18)$$

4.3 Evolutionary Game Formulation and Analysis

In this section, we model the dynamic AP and service selection of the STs in the backscatter-assisted RF-powered CR network by evolutionary game, and reach the evolutionary equilibrium. In our work, the evolutionary equilibrium means that all the STs have no motivation to change their AP and service selection strategy, and their AP and service selection strategy remains unchanged. Therefore, the evolutionary equilibrium is also the Nash equilibrium in traditional non-cooperative game [5]. In particular, we first present the formulation of the evolutionary game. We then prove important properties of the formulated evolutionary game, i.e., the existence and uniqueness, and the stability of the evolutionary equilibrium. Furthermore, we consider the delay of the information used by the STs to adapt their selection in the replicator dynamics. We provide some analyses about the effect of the delay on the stability of the evolutionary equilibrium.

4.3.1 Evolutionary Game Formulation

We employ the evolutionary game to model the dynamic AP and service selection of the STs in the network, and the players, the population, the strategy, and the payoff are described as follows.

- *Players*: The STs compete for the time resource for data transmission. Therefore, each ST is a player and we have N players in the network.
- *Population*: The STs form a population in our formulated game. As such, we have one population in the network.
- *Strategy*: The strategy of each ST is the selection of an AP and a service. Accordingly, the set of the strategies of each ST includes $3M$ elements.
- *Payoff*: The payoff of each ST is its utility defined in (4.18). The more the transmitted data per unit price, the higher the payoff.

Since $x_{m,i}$ varies over time in the evolutionary game, we denote $x_{m,i}(t)$ as the value of $x_{m,i}$ at time t. Accordingly, we denote $u_{m,i}(t)$ as the value of $u_{m,i}$ at time t. Therefore, we take $3M$ selection strategies into account, and the expected utility of one ST at time t is

$$\bar{u}(t) = \sum_{m \in M} \sum_{i=1}^{3} x_{m,i}(t) u_{m,i}(t). \tag{4.19}$$

Then, we employ the replicator dynamics to model the AP and service adaptation of the STs. Specifically, the replicator dynamic process of the STs is formulated as a series of ordinary differential equations (ODEs), as follows:

$$\dot{x}_{m,i}(t) = f_{m,i}(\mathbf{x}(t)) = \mu x_{m,i}(t)(u_{m,i}(t) - \bar{u}(t)),$$
$$\forall m \in M, \forall i \in \{1, 2, 3\}, \forall t, \tag{4.20}$$

with the initial state $\mathbf{x}(0) \in X$, where $\mathbf{x}(t) = [x_{1,1}(t), x_{1,2}(t), x_{1,3}(t), \ldots, x_{M,3}(t)]$, $\dot{x}_{m,i}(t)$ denotes the time derivative of $x_{m,i}(t)$ with respect to t, X is the set of all feasible states, and μ is a positive learning rate, which controls the speed of the strategy adaptation. From (4.19) and (4.20), if $\sum_{m \in M} \sum_{i=1}^{3} x_{m,i}(0) = 1$, we have

$$\sum_{m \in M} \sum_{i=1}^{3} \dot{x}_{m,i}(t) = 0, \forall t, \tag{4.21}$$

and

$$\sum_{m \in M} \sum_{i=1}^{3} x_{m,i}(t) = 1, \forall t. \tag{4.22}$$

4.3.2   Existence and Uniqueness of the Evolutionary Equilibrium

To prove the existence and uniqueness of the evolutionary equilibrium, we first present the following lemma.

Lemma 4.1 $f_{m,i}(\mathbf{x}(t))$, $\forall \mathbf{x}(t) \in X$, $\forall m \in M$, $\forall i \in \{1, 2, 3\}$, is bounded.

Proof From (4.1), (4.6), and (4.18), we have

$$x_{m,1}(t)u_{m,1}(t) = \frac{\beta_m^{\text{idle}} \gamma_m^{\text{idle}} \xi B}{N p_{m,1}} \log_2 \left(1 + \frac{\varsigma_m \alpha x_{m,1}(t) N}{\beta_m^{\text{idle}} \gamma_m^{\text{idle}}} \right), \tag{4.23}$$

from which we can easily conclude that $x_{m,1}(t)u_{m,1}(t)$ is bounded.

From (4.2), (4.3), (4.12), and (4.18), we have

$$h_{m,\min}(t) \le x_{m,2}(t)u_{m,2}(t) \le h_{m,\max}(t), \tag{4.24}$$

where

$$h_{m,\min}(t) = \frac{(1 - \beta_m^{\text{idle}}) \gamma_m^{\text{idle}} \xi B}{N p_{m,2}} \log_2 \left(1 + \frac{\varsigma_m \alpha x_{m,2}(t) N}{(1 - \beta_m^{\text{idle}}) \gamma_m^{\text{idle}}} \right), \tag{4.25}$$

and

$$h_{m,\max}(t) = \frac{R_m^{\text{b}} (1 - \beta_m^{\text{busy}}) \gamma_m^{\text{busy}}}{N p_{m,2}} + h_{m,\min}(t). \tag{4.26}$$

Here, $h_{m,\min}(t)$ is obtained when $\tau_{m,2}^{\text{b}} = 0$ in (4.12), and $h_{m,\max}(t)$ is obtained when we ignore the reduction of the harvested energy due to the backscatter communication. From (4.25) and (4.26), we can conclude that both $h_{m,\min}(t)$ and $h_{m,\max}(t)$ are bounded. Therefore, from (4.24), $x_{m,2}(t)u_{m,2}(t)$ is bounded.

From (4.4), (4.17), and (4.18), we have

$$x_{m,3}(t)u_{m,3}(t) = \frac{R_m^{\text{b}} \beta_m^{\text{busy}} \gamma_m^{\text{busy}}}{N p_{m,3}}, \tag{4.27}$$

from which, we have that $x_{m,3}(t)u_{m,3}(t)$ is bounded. Since $x_{m,i}(t)u_{m,i}(t)$, $\forall m \in M$, $\forall i \in \{1, 2, 3\}$, is bounded, from (4.19), $\bar{u}(t)$ is bounded, from which we have that $x_{m,i}(t)\bar{u}(t)$ is also bounded. Therefore, from (4.20), $f_{m,i}(\mathbf{x}(t))$, $\forall \mathbf{x}(t) \in X$, $\forall m \in M$, $\forall i \in \{1, 2, 3\}$, is bounded.   ∎

Lemma 4.1 paves the way for proving the existence and uniqueness of the evolutionary equilibrium, which is given in the following theorem.

Theorem 4.1 *For any initial state* $\mathbf{x}(0) \in X$, *the evolutionary game defined by (4.20) is solvable, and admits a unique evolutionary equilibrium.*

Proof From Lemma 4.1, we have that $f_{m,i}(\mathbf{x}(t))$, $\forall \mathbf{x}(t) \in X$, $\forall m \in M$, $\forall i \in \{1, 2, 3\}$, is bounded. In addition, from (4.18) and (4.20), we find that $f_{m,i}(\mathbf{x}(t))$, $\forall \mathbf{x}(t) \in X$, is continuous. Therefore, there exists a positive parameter Q, such that

$$\left| f_{m,i}(\mathbf{x}'(t)) - f_{m,i}(\mathbf{x}''(t)) \right| \leq Q \left| \mathbf{x}'(t) - \mathbf{x}''(t) \right|,$$
$$\forall \mathbf{x}'(t), \mathbf{x}''(t) \in X, \forall m \in M, \forall i \in \{1, 2, 3\}, \forall t, \qquad (4.28)$$

which indicates that $f_{m,i}(\mathbf{x}(t))$ satisfies the global Lipschitz condition. From the Cauchy–Lipschitz theorem [12], it follows that for any initial state $\mathbf{x}(0) \in X$, the evolutionary game defined by (4.20) is solvable, and admits a unique evolutionary equilibrium. ∎

From Theorem 4.1, we can conclude that a unique evolutionary equilibrium can be eventually reached in our formulated evolutionary game.

4.3.3 Stability Analysis of the Evolutionary Equilibrium

We now prove that the evolutionary equilibrium is stable, which is shown in the following theorem.

Theorem 4.2 *For any initial state* $\mathbf{x}(0) \in X$, *the evolutionary game defined by (4.20) admits a stable evolutionary equilibrium.*

Proof Define a Lyapunov function as

$$L(\mathbf{x}(t)) = \left(\sum_{m \in M} \sum_{i=1}^{3} x_{m,i}(t) \right)^2. \qquad (4.29)$$

From (4.29), we have

$$L(\mathbf{x}(t)) \begin{cases} = 0 & \text{if } \mathbf{x}(t) = \mathbf{0}, \\ > 0 & \text{otherwise.} \end{cases} \qquad (4.30)$$

In addition, obtaining the first-order derivative of $L(x(t))$ with respect to t, we have

$$\frac{d(L(\mathbf{x}(t)))}{dt} = 2 \left(\sum_{m \in M} \sum_{i=1}^{3} x_{m,i}(t) \right) \left(\sum_{m \in M} \sum_{i=1}^{3} \dot{x}_{m,i}(t) \right). \qquad (4.31)$$

From (4.21), we have

$$\frac{d(L(\mathbf{x}(t)))}{dt} = 0. \tag{4.32}$$

Therefore, $L(\mathbf{x}(t))$ satisfies the required conditions of the Lyapunov function, which is defined in the Lyapunov's second method for stability [13]. Accordingly, it immediately follows that for any initial state $x(0) \in X$, the evolutionary game defined by (4.20) admits a stable evolutionary equilibrium. ∎

From Theorem 4.2, we can conclude that the obtained evolutionary equilibrium of our formulated evolutionary game is stable.

4.3.4 Delay in Replicator Dynamics

In practice, the calculation of the expected payoff at the controller (e.g., an IoT application server) and the information exchange between the controller and the STs may cause the delay of the information at the STs. Therefore, the STs may not adapt their strategies according to the current expected payoff. In this case, the STs have to adapt their strategies based on the historical information. Specifically, each ST adapts the AP and the service selection at time t based on the state information at time $t - \delta$, where δ is the delay of the information used by the STs. Accordingly, the delayed replicator dynamic process can be expressed as

$$\dot{x}_{m,i}(t) = \mu x_{m,i}(t - \delta)(u_{m,i}(t - \delta) - \bar{u}(t - \delta)),$$
$$\forall m \in \mathcal{M}, \forall i \in \{1, 2, 3\}. \tag{4.33}$$

From (4.33), we find that each ST needs the information at time $t = -\delta$ when making a decision on the AP and service selection at time $t = 0$. For the convenience of the analysis, we assume that the STs have the information at time $t = 0$ when the STs choose APs and services at time $t < \delta$ [6]. As a result, each ST has the available information when making a decision at any time by using the ODEs in (4.33).

When there exists a delay in the replicator dynamics, the stability of the evolutionary equilibrium cannot always be guaranteed. According to [7], for a small delay, the evolutionary game with the delayed replicator dynamics will converge to the evolutionary equilibrium of the evolutionary game with the replicator dynamics in (4.20). However, when the delay is larger than a certain threshold, the evolutionary game with the delayed replicator dynamics will not converge and the stability can not be guaranteed. Due to the high nonlinearity of the utility functions, it is challenging, if not impossible, to analytically derive the delay threshold to guarantee the stability. However, we can resort to numerical simulations to evaluate the stability of the delayed replicator dynamics, as shown in Sect. 4.6. Nevertheless, we can analytically derive the stability region of the delayed replicator dynamics in a special case in Sect. 4.4.

4.4 Stability Region of Delayed Replicator Dynamics in a Special Case

In this section, we analytically derive the stability region of the delayed replicator dynamics in a special case.

4.4.1 Descriptions of the Special Case

In the special case, we consider a secondary network where the primary channel is always busy [4]. In this case, only backscatter mode can be adopted for the STs transmitting data to the APs. Therefore, only Service 3 can be provided by each AP and we only consider the AP selection in this case. Let N_m denote the number of the STs choosing AP m. Accordingly, the proportion of the STs choosing AP m is $x_m = N_m/N$. Therefore, the assigned time period for each ST choosing AP m is

$$\tau_m^{\text{busy}} = \frac{\gamma_m^{\text{busy}}}{x_m N}. \tag{4.34}$$

Accordingly, the amount of the transmitted data of each ST choosing AP m is

$$Z_m = R_m^{\text{b}} \tau_m^{\text{busy}} = \frac{R_m^{\text{b}} \gamma_m^{\text{busy}}}{x_m N}. \tag{4.35}$$

Let p_m denote the price of each ST choosing AP m. Therefore, the utility of each ST choosing AP m is

$$u_m = \frac{Z_m}{p_m} = \frac{R_m^{\text{b}} \gamma_m^{\text{busy}}}{x_m N p_m}. \tag{4.36}$$

Accordingly, the expected utility of each ST is

$$\bar{u}^{\text{s}} = \sum_{m \in M} x_m u_m. \tag{4.37}$$

From (4.36) and (4.37), we can formulate an evolutionary game to model the AP selection of the STs by replicator dynamics. Since our considered network in this section is a special case of that in Sect. 4.2, the formulated evolutionary game in Sect. 4.3 can be easily adapted to the network in this section. Furthermore, the existence and uniqueness of the evolutionary equilibrium are guaranteed. We can also evaluate the stability of the delayed replicator dynamics by numerical simulations in this special case. We can further analytically derive stability region of the delayed replicator dynamics in this case, which is detailed in the following.

4.4.2 Stability Region of the Delayed Replicator Dynamics

Let $x_m(t)$, $u_m(t)$, and $\bar{u}^s(t)$ denote the values of x_m, u_m, and $\bar{u}^s$, respectively, at time t. In this special case, the delayed replicator dynamics can be formulated as

$$\dot{x}_m(t) = \mu x_m(t - \delta)(u_m(t - \delta) - \bar{u}^s(t - \delta)), \forall m \in \mathcal{M}, \tag{4.38}$$

where $\dot{x}_m(t)$ is the time derivative of $x_m(t)$ with respect to t. We now present the stability region of the evolutionary equilibrium with the delayed replicator dynamics in the following theorem.

Theorem 4.3 *The stability of the evolutionary equilibrium with the delayed replicator dynamics in (4.38) can be guaranteed if and only if*

$$\delta < \frac{\pi}{2\mu \sum_{m \in \mathcal{M}} \frac{R_m^b \gamma_m^{busy}}{N p_m}}. \tag{4.39}$$

Proof We can equivalently rewrite the delayed replicator dynamics in (4.38) as

$$\dot{\mathbf{x}}^s(t) = \mathbf{A}\mathbf{x}^s(t - \delta) + \mathbf{c}, \tag{4.40}$$

where

$$\dot{\mathbf{x}}^s(t) = [\dot{x}_1(t), \dot{x}_2(t), \dots, \dot{x}_M(t)]^T,$$

$$\mathbf{x}^s(t - \delta) = [x_1(t - \delta), x_2(t - \delta), \dots, x_M(t - \delta)]^T,$$

$$\mathbf{c} = \left[\frac{\mu R_1^b \gamma_1^{busy}}{N p_1}, \frac{\mu R_2^b \gamma_2^{busy}}{N p_2}, \dots, \frac{\mu R_M^b \gamma_M^{busy}}{N p_M} \right]^T,$$

and

$$\mathbf{A} = -\kappa \mathbf{I}.$$

Here, $\mathbf{I}$ is an M dimensional unit matrix, and κ is defined as

$$\kappa = \mu \sum_{m \in \mathcal{M}} \frac{R_m^b \gamma_m^{busy}}{N p_m}. \tag{4.41}$$

From (4.40), the characteristic equation is given by

$$\lambda + \kappa \exp(-\lambda \delta) = 0. \tag{4.42}$$

According to [14], the evolutionary equilibrium with the delayed replicator dynamics in (4.38) is stable if and only if the real parts of all the roots are negative. This condi-

Algorithm 4.1 Algorithm of the AP and Service Selection

1: Initialize iteration counter $j \leftarrow 0$, the maximum number of iterations $j_{\max}$, and the constant to control the speed of the adaptation ρ.
2: Each ST randomly chooses an AP and its service, and set iteration counter $j \leftarrow 0$.
3: **while** $j < j_{\max}$ **do**
4:   $j \leftarrow j + 1$
5:   Each ST calculates the utility (i.e., $u_{m,i}$) based on the assigned time resource and then sends the utility information to a controller.
6:   The controller calculates the expected utility of the STs (i.e., $\bar{u}$), and sends the expected utility and the set of selection strategies with $u_{m,i} > \bar{u}$ to each ST.
7:   **for** all the STs with $u_{m,i} < \bar{u}$ **do**
8:     **if** $randn() < \frac{\rho(\bar{u}-u_{m,i})}{\bar{u}}$ **then**
9:       Randomly choose a service provided by the APs with the utility higher than $\bar{u}$.
10:     **end if**
11:   **end for**
12: **end while**

tion further needs $2\kappa\delta < \pi$. Therefore, the stability region of the delayed replicator dynamics in (4.38) is

$$\delta < \frac{\pi}{2\kappa} = \frac{\pi}{2\mu \sum_{m \in M} \frac{R_m^b \gamma_m^{busy}}{N p_m}}, \tag{4.43}$$

which completes the proof of Theorem 4.3. ∎

According to Theorem 4.3, when the delay in the replicator dynamics is smaller than the threshold (i.e., $\pi/(2\kappa)$), the evolutionary game will converge to the evolutionary equilibrium solution. Otherwise, the evolutionary game cannot converge to the evolutionary equilibrium solution [15].

4.5 Algorithm of the AP and Service Selection

In this section, we develop an algorithm to implement the AP and service selection in the backscatter-assisted RF-powered CR network based on evolutionary game.

Considering that the PT transmits data on a periodic basis, each ST adapts its AP and service selection at the beginning of each time unit. Therefore, the STs adapt their strategies per time unit, i.e., one period of the primary signal. The overall process of the AP and service selection in the network is summarized in Algorithm 4.1. Each ST randomly chooses a service provided by the APs and calculates its utility, which is sent to the controller, such as an IoT application server. The controller calculates the expected utility and then broadcasts the expected utility and the set of selection strategies with the utilities higher than the expected utility to all the STs. Some of the STs with the utilities lower than the expected utility will adapt their selection and randomly choose the AP and service strategy with the utility higher than the expected utility, as shown from Line 7 to Line 11. Therefore, the strategies with the utilities lower than the expected utility will be chosen by fewer STs, and the utilities of the STs choosing these strategies will increase. Similarly, the strategies with the utilities higher than the expected utility will be chosen by more STs, and the utilities of the STs choosing these strategies will decrease. Note that although the algorithm needs time for the STs adapting their strategies, the time period for the strategy adaptation is generally acceptable. The reason is that the STs still transmit data to the APs during the strategy adaptation period, although the utility of one ST choosing the same AP and service strategy may differ between the last and the current time units. When all the STs choosing any strategy can achieve almost the same utilities, the STs almost do not change their selection strategies. This process is very similar to that in the replicator dynamics. In addition, the communication between the controller and the STs may cause the delay of the information used by the STs. In the jth iteration, when the STs adapt their strategies by using the information (including the expected utility and the set of strategies with $u_{m,i} > \bar{u}$) calculated from the assigned time resource in the $(j-1)$th iteration, the delay is regarded as 0. Accordingly, when the STs use the information calculated from the assigned time resource in the $(j-\delta)$th iteration to adapt their strategy, the delay is regard as $\delta - 1$.

We now evaluate the computational complexity of our developed algorithm. From the implementation process of the STs choosing the APs and the services, we find that each ST only calculates its own utility, sends its utility to the controller, compares its utility with the expected utility, and adapts its selection if its utility is below the expected utility. Therefore, the complexity of each ST does not increase with the numbers of STs and APs, and the complexity of the controller linearly increases with the number of STs. Accordingly, the complexities of each ST and the controller are $O(1)$ and $O(N)$, respectively. This result indicates that the developed algorithm is computationally efficient and also scalable. In overall, the computational complexity of implementing the AP and service selection based on the evolutionary game is very low, which is suitable for the backscatter-assisted RF-powered CR network.

4.6 Numerical Results

In this section, we present the numerical results to demonstrate the effectiveness of the developed dynamic AP and service selection in the backscatter-assisted RF-powered CR network. We consider a network with a PT, 100 STs, and 2 APs, and the channel gains from the STs to AP 1 and AP 2 are -55 dB and -45 dB, respectively. The normalized busy period is 0.8, which is equally available for each AP in the TDMA manner. Similarly, the normalized idle period is also equally available for each AP in the TDMA manner. The fractions of the idle periods for Service 1 and the fractions of the busy periods for Service 3 at 2 APs are all 0.5. Considering the fact that a service with more time resource and a better channel condition is usually marked with a higher price, the prices of Service 1, Service 2, and Service 3 provided by AP 1 are 0.3, and 0.5, and 0.1, respectively, and the prices of Service 1, Service 2,

Fig. 4.3 Proportions of the STs choosing different APs and services over time

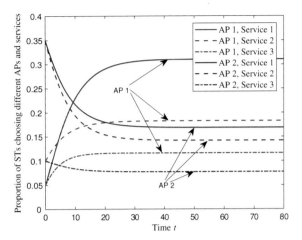

Fig. 4.4 Utilities of the STs choosing different APs and services over time

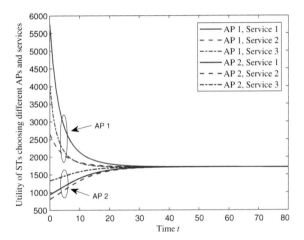

and Service 3 provided by AP 2 are 0.8, 1, and 0.15, respectively. The bandwidth of the channel is 0.1 MHz, and the transmission rates in the backscatter mode are all 10 kbps [4]. The initial state of the STs $\mathbf{x}(0)$ is $[0.05, 0.1, 0.05, 0.35, 0.35, 0.1]$, and the learning rate μ is 10^{-4}.

We first investigate the proportions and the utilities of the STs versus the evolutionary time. Figures 4.3 and 4.4 plot the proportions and the utilities of the STs choosing different APs and services, respectively. From these figures, we find that the proportions of the STs choosing different APs and services first change and then remain unchanged over the evolutionary time. Accordingly, the utilities of the STs choosing different APs and services vary until the evolutionary equilibrium is reached. We also find that when the evolutionary equilibrium is reached, the STs can achieve the same utilities although they choose different APs and services. This is due to the fact that the evolutionary equilibrium is reached only when the utilities of

Fig. 4.5 Phase plane of
replicator dynamics

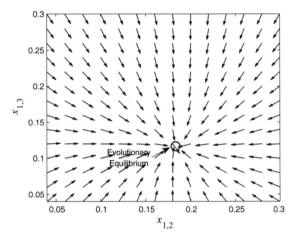

Fig. 4.6 Time to reach the
evolutionary equilibrium
versus μ when $N = 100$,
150, or 200

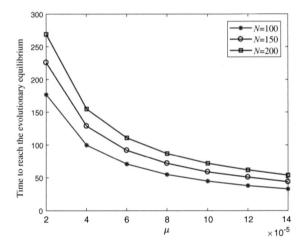

the STs choosing any service provided by any AP are equal to the expected utility
of the STs.

We next examine the dynamic behaviors of the STs in Fig. 4.5. For an ease of
presentation, we illustrate the phase plane of the replicator dynamics of $x_{1,2}$ and $x_{1,3}$.
In this case, we assume that the proportions of the STs choosing services provided
by AP 2 (i.e., $x_{2,1}$, $x_{2,2}$, and $x_{2,3}$) can achieve the equilibrium, and $x_{1,1} = 1 - x_{1,2} -
x_{1,3} - x_{2,1} - x_{2,2} - x_{2,3}$. In the phase plane, the STs follow the directions of the
arrows to adapt their strategies. From this figure, we observe that any initial state
will eventually reach the evolutionary equilibrium. This result also verifies that our
obtained evolutionary equilibrium is stable, which is in accordance with Theorem 4.2.

We now discuss the impact of the learning rate μ on the AP and service selection
of the STs. Figure 4.6 plots the time to reach the evolutionary equilibrium versus
μ for the numbers of STs $N = 100$, 150, or 200. From this figure, we find that the

Fig. 4.7 Proportion of the
STs choosing Service 3
provided by AP 1 (i.e., $x_{1,3}$)
over evolutionary time with
different delays

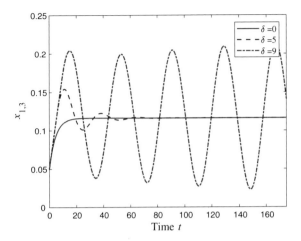

value of μ influences the speed of achieving the evolutionary equilibrium. The larger
the value of μ is, the faster the evolutionary equilibrium is reached. In addition, we
observe that the number of the STs N also has an impact on the time to reach the
evolutionary equilibrium. With a given value of μ, the larger the value of N is, the
more time is needed to reach the evolutionary equilibrium.

We now evaluate the impact of delay δ on the proportions and the utilities of
the STs choosing different APs and services. We take the STs choosing Service 3
provided by AP 1 as an example. Figure 4.7 plots the proportion of the STs choosing
Service 3 provided by AP 1 (i.e., $x_{1,3}$) over the evolutionary time when $\delta = 0, 5$,
and 9, respectively. From this figure, we find that $x_{1,3}$ can reach the evolutionary
equilibrium when $\delta = 5$, while $x_{1,3}$ fluctuates over time when $\delta = 9$. When the
value of δ is relatively small, the evolutionary equilibrium can be reached. When
the value of δ becomes larger, the obtained information becomes more inaccurate,
and the equilibrium is never reached. We also observe that the reached evolutionary
equilibriums are the same when $\delta = 0$ (i.e., without delay) and $\delta = 5$. This result
shows that a relatively small delay does not affect the proportions of the STs choosing
different APs and services when the evolutionary equilibrium is reached. In addition,
compared with the dynamics when $\delta = 0$, we find that there are more fluctuations,
and it takes longer time to reach the evolutionary equilibrium when $\delta = 5$. The
larger the delay is, the slower the equilibrium is reached. Figure 4.8 plots the utility
of the STs choosing Service 3 provided by AP 1 (i.e., $u_{1,3}$) over the evolutionary
time when $\delta = 0, 5$, and 9, respectively. This figure shows that the AP and service
selection strategies can eventually achieve the same utility of the STs when $\delta = 0$
or $\delta = 5$, and the utility of the STs fluctuates over time when $\delta = 9$. This result is
consistent with that in Fig. 4.7.

We now investigate the impact of δ on the proportions and the utilities of the STs
choosing different APs in the special case under our consideration, and verify the
effectiveness of Theorem 4.3. In Figs. 4.9 and 4.10, we plot the proportion and the

Fig. 4.8 Utility of the STs choosing Service 3 provided by AP 1 (i.e., $u_{1,3}$) over evolutionary time with different delays

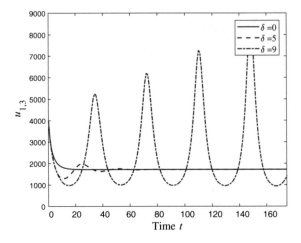

Fig. 4.9 Proportion of the STs choosing AP 1 (i.e., x_1) over evolutionary time with different delays in the special case

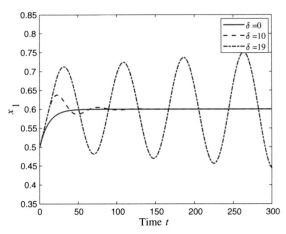

Fig. 4.10 Utility of the STs choosing AP 1 (i.e., u_1) over evolutionary time with different delays in the special case

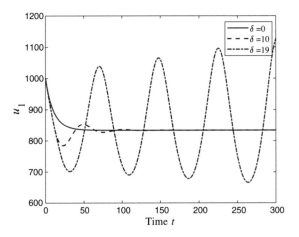

Fig. 4.11 Proportions of the STs choosing different APs and services versus the price of Service 1 provided by AP 1 (i.e., $p_{1,1}$)

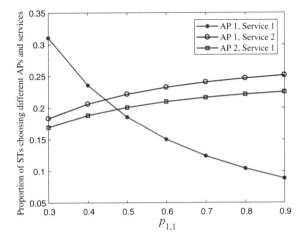

utility of the STs choosing AP 1 (i.e., x_1 and u_1), respectively, over the evolutionary time with different delays in the special case. From these figures, we observe that the same evolutionary equilibrium is reached when $\delta = 0$ and $\delta = 10$, while x_1 and u_1 always vary over time when $\delta = 19$. This result is similar to that in Figs. 4.7 and 4.8. Furthermore, we can verify that $\delta = 10$ satisfies (4.39) in Theorem 4.3, but $\delta = 19$ violates it. This demonstrates the effectiveness of Theorem 4.3.

We now discuss the impact of prices on the obtained evolutionary equilibrium. Figures 4.11 and 4.12 plot the proportions of the STs choosing different APs and services and the utilities of the STs, respectively, versus the price of Service 1 provided by AP 1 (i.e., $p_{1,1}$) when the evolutionary equilibrium is reached. When $p_{1,1}$ increases, the proportion of the STs choosing Service 1 provided by AP 1 (i.e., $x_{1,1}$) decreases, and the proportions of the STs choosing other APs and/or services rise. A larger $p_{1,1}$ leads to a lower utility of the STs choosing Service 1 provided by AP 1. As such, the STs that choose Service 1 provided by AP 1 have less benefit to choose this service, and some of the STs will switch to other APs and/or services instead. This process will increase the utility of the STs choosing Service 1 provided by AP 1 and reduce that of the STs choosing other APs and/or services. When all the utilities of the STs choosing different APs and services are the same, the evolutionary equilibrium is reached with the new value of $p_{1,1}$. Therefore, the utilities of the STs decrease when $p_{1,1}$ increases, as shown in Fig. 4.12. From this figure, we also find that a larger transmission rate in the backscatter mode R^b leads to a higher utility of the STs when the evolutionary equilibrium is reached. A larger R^b can bring a higher throughput of the STs. Therefore, the utilities of the STs increase with R^b.

We now focus on the impact of the number of STs N on the utilities of the STs when the evolutionary equilibrium is reached. In Fig. 4.13, we plot the utilities of the STs versus N when $R^b = 5$ kbps, 10 kbps, or 20 kbps. This figure shows that N has a considerable impact on the utilities of the STs. We observe that the utilities of the STs decrease more than 46% when N increases from 100 to 220. With more STs

Fig. 4.12 Utilities of the
STs versus the price of
Service 1 provided by AP 1
(i.e., $p_{1,1}$) when the
transmission rate in the
backscatter mode
$R^b = 5$ kbps, 10 kbps, or
20 kbps

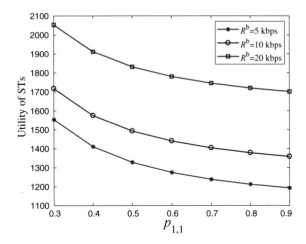

Fig. 4.13 Utilities of the
STs versus the number of the
STs N when the transmission
rate in the backscatter mode
$R^b = 5$ kbps, 10 kbps, or
20 kbps

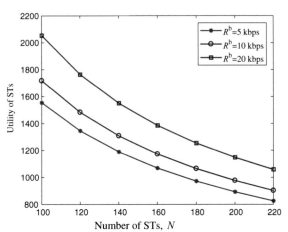

connected to the network, the time resource assigned to each ST is reduced, which
results in a lower throughput of the STs and hence a lower utility of the STs. We
further observe that a larger R^b will lead to a higher utility, which is similar to the
result in Fig. 4.12.

We finally evaluate the performance of the developed algorithm of the AP and
service selection in the network. Figure 4.14 plots the utilities of the STs choosing
different APs and services over time by using the algorithm. From this figure, we
observe that the utilities of the STs choosing different APs and services first adjust
and then keep almost unchanged, and the utilities of the STs choosing different APs
and services are finally almost the same, which is similar to the result in Fig. 4.4.
Figure 4.15 plots the utility of the STs choosing Service 3 provided by AP 1 (i.e.,
$u_{1,3}$) over time with different delays by using the algorithm. We observe that with
a relatively small delay, $u_{1,3}$ can still converge to the equilibrium which is reached

Fig. 4.14 Utilities of the STs choosing different APs and services over time by using the algorithm

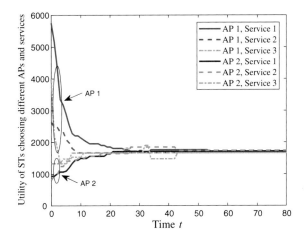

Fig. 4.15 Utility of the STs choosing Service 3 provided by AP 1 over time with different delays by using the algorithm

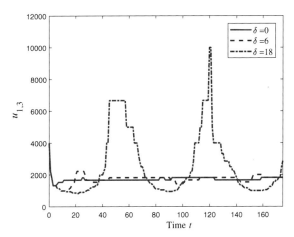

with no delay of the information. When the delay is relatively large, $u_{1,3}$ fluctuates over time and cannot converge, which is similar to the result in Fig. 4.8. The results in Figs. 4.14 and 4.15 demonstrate the effectiveness of the developed algorithm, and show that the results obtained from the replicator dynamics can provide useful insights for the algorithm of the AP and service selections in the backscatter-assisted RF-powered CR network.

4.7   Conclusions

In this chapter, we have investigated the dynamic AP and service selection in a backscatter-assisted RF-powered CR network. Specifically, we have formulated the joint AP and service selection problem as an evolutionary game. In the evolutionary game, the adaptation of the AP and service selection of the STs has been modelled as the replicator dynamics. We have proved that there exists a unique evolutionary equilibrium in our formulated game, and the evolutionary equilibrium is stable. Also, we have considered the replicator dynamics with delay of the information, and have shown that an evolutionary equilibrium can be also reached when the delay is relatively small. The stability region of the delayed replicator dynamics has been analytically derived for a special case. Furthermore, we have developed an algorithm to implement the AP and service selection in the network, which can obtain the similar results by using the replicator dynamics.

References

1. V. Liu, A. Parks, V. Talla, S. Gollakota, D. Wetherall, J.R. Smith, Ambient backscatter: wireless communication out of thin air, in *ACM SIGCOMM* (ACM, 2013), pp. 39–50
2. H. Ju, R. Zhang, Throughput maximization in wireless powered communication networks. IEEE Trans. Wireless Commun. **13**(1), 418–428 (2014)
3. D.T. Hoang, D. Niyato, P. Wang, D.I. Kim, Z. Han, The tradeoff analysis in RF-powered backscatter cognitive radio networks, in *IEEE Globecom* (IEEE, 2016), pp. 1–6
4. D.T. Hoang, D. Niyato, P. Wang, D.I. Kim, Z. Han, Ambient backscatter: a new approach to improve network performance for RF-powered cognitive radio networks. IEEE Trans. Commun. **65**(9), 3659–3674 (2017)
5. Z. Han, D. Niyato, W. Saad, T. Basar, A. Hjorungnes, *Game Theory in Wireless and Communication Networks: Theory, Models, and Applications* (Cambridge University Press, Cambridge, 2011)
6. D. Niyato, E. Hossain, Dynamics of network selection in heterogeneous wireless networks: an evolutionary game approach. IEEE Trans. Veh. Technol. **58**(4), 2008–2017 (2009)
7. P. Semasinghe, E. Hossain, K. Zhu, An evolutionary game for distributed resource allocation in self-organizing small cells. IEEE Trans. Mobile Comput. **14**(2), 274–287 (2015)
8. J. Hofbauer, K. Sigmund, Evolutionary game dynamics. Bull. Amer. Math. Soc. **40**(4), 479–519 (2003)
9. K. Zhu, E. Hossain, D. Niyato, Pricing, spectrum sharing, and service selection in two-tier small cell networks: a hierarchical dynamic game approach. IEEE Trans. Mobile Comput. **13**(8), 1843–1856 (2014)
10. X. Wang, F.-C. Zheng, P. Zhu, X. You, Energy-efficient resource allocation in coordinated downlink multi-cell OFDMA systems. IEEE Trans. Veh. Technol. **65**(3), 1395–1408 (2016)
11. Y. Zhang, J. An, K. Yang, X. Gao, J. Wu, Energy-efficient user scheduling and power control for multi-cell OFDMA networks based on channel distribution information. IEEE Trans. Singal Process. **66**(22), 5841–5861 (2018)
12. J.C. Engwerda, *LQ Dynamic Optimization and Differential Games* (Wiley, New York, 2005)
13. S. Sastry, Lyapunov stability theory. Nonlinear systems (1999), pp. 182–234
14. K. Gopalsamy, *Stabiliby and Oscillation in Delay Differential Equations of Population Dynamics* (Springer, Berlin, 1992)
15. H. Tembine, E. Altman, R. El-Azouzi, Y. Hayel, Bio-inspired delayed evolutionary game dynamics with networking applications. Telecommun. Syst. **47**(1), 137–152 (2011)

Chapter 5
Throughput-Maximized Relay Mode Selection and Resource Sharing

In this chapter, we investigate the throughput-maximized relay mode selection and resource sharing for backscatter-assisted hybrid relay networks. Section 5.1 introduces the background of developing throughput-maximized relay selection and resource sharing scheme and summarizes the contributions. Section 5.2 presents the system model, and Sect. 5.3 formulates the throughput maximization problem. Section 5.4 develops the algorithm to derive the solution to the formulated throughput maximization problem. Section 5.5 presents the simulation results, and Sect. 5.6 concludes this chapter.

5.1 Introduction

Recently, wireless backscatter has been introduced as a new promising communication technique that is featured with extremely low power consumption by transmitting information in the passive mode via the modulation and reflection of the ambient RF signals [1, 2]. The architectural differences between the active and passive radios lead to complement transmission capabilities and power demands in two modes, e.g., [3, 4]. This implies that a hybrid wireless system with both active and passive radios can flexible schedule the data transmissions in different modes according to the channel conditions, energy status, and traffic demands, and thus achieve a higher network performance, especially for energy constrained IoT networks.

Due to the extremely low power consumption of the passive radios, it becomes promising to use the passive relays to assist the active RF communications, especially for wirelessly powered IoT networks with energy harvesting constraints, e.g., [5–7]. In this chapter, we develop a hybrid relay communications model, in which both the passive and active relays are employed simultaneously to assist the active RF communications. Each relay can switch between the active and passive modes independently to maximize the overall throughput, according to its channel conditions and energy status.

Specifically, we focus on a two-hop hybrid relay communication model in this chapter. In the first hop, the multi-antenna transmitter beamforms information to the relays and the receiver. The set of passive relays instantly backscatter the RF signals to enhance signal reception at both of the active relays and the receiver. In the second hop, the active relays jointly beamform the received signals to the receiver. Meanwhile, the passive relays can adapt their reflection coefficients to enhance the active relays' forwarding channels. We aim to maximize the overall network throughput by jointly optimizing the transmit beamforming, the relays' mode selection, and their operating parameters. It is noticed that the throughput maximization problem is combinatorial and difficult to solve optimally. To overcome this difficulty, we develop a two-step solution to optimize the relay strategy. Specifically, with a fixed relay mode, we firstly find lower bounds on the signal-to-noise ratio (SNR) at the receiver under different channel conditions, based on which we devise a set of performance metrics to evaluate individual relay's performance gain. Then, we provide an iterative procedure to update the relays' mode selection to improve the overall relay performance. The main contributions of this chapter are summarized as follows:

1. Different from the conventional relay communications, multiple energy harvesting relays in both the active and passive modes are employed to collaborate in relay communications. The active relays follow the AF protocol while the passive relays can adapt their reflection coefficients to enhance the relay channels.
2. Two lower bounds are derived to evaluate the receiver's SNR under different channel conditions. The SNR evaluation serves as a performance metric for relay mode selection. Each lower bound requires to solve an optimization problem involving the transmit beamforming, the active relays' power control, and the passive relays' phase control.
3. To bypass the complexity in SNR evaluation, we also derive a set of heuristic algorithms based on simple approximations of the SNR performance to evaluate each passive relay's performance gain. The simulation results verify that the SNR-based mode selection can achieve the optimal throughput performance, while the heuristic algorithms also perform well with significantly reduced complexity.

5.2 System Model

We consider a multi-input single-ouput (MISO) downlink communication system with a group of single-antenna user devices coordinated by a multi-antenna HAP. The HAP's information transmissions to different user devices can be scheduled in orthogonal channels, e.g., via a TDMA protocol [6]. Without loss of generality, we focus on the simplest case with only one receiver. The user devices can serve as the wireless relays for each other via device-to-device (D2D) communications. In particular, the data transmission from the HAP to the receiver can be assisted by a set of relays following the AF protocol. The set of relays is denoted by $\mathcal{N} = \{1, 2, \ldots, N\}$. The HAP has a constant power supply, while the relays are wirelessly powered by RF signals emitted from the HAP. Via signal beamforming, the HAP can control the information rate and power transfer to the relays following the power-splitting (PS) protocol [8]. Note that the relay optimization with the time-switching (TS) protocol has a similar structure with the PS protocol [9]. Hence, we focus on the PS protocol in this work. Assuming that the HAP has K antennas, let $\mathbf{f}_0 \in \mathbb{C}^K$ and $\mathbf{f}_n \in \mathbb{C}^K$ denote the complex channels from the HAP to the receiver and from the HAP to the nth relay, respectively. Let $\mathbf{g} \triangleq [g_1, g_2, \ldots, g_N]^T \in \mathbb{C}^N$ denote the complex channels from the relays to the receiver. All the channels are assumed to be block fading and can be estimated in a training period before data transmissions [10].

5.2.1 Two-Hop Hybrid Relaying Scheme

The relay-assisted information transmission follows a two-hop half-duplex protocol. As shown in Fig. 5.1, the information transmission is divided into two phases, i.e., the relay receiving and forwarding phases, corresponding to the information transmission in two hops. Due to a short distance between transceivers in a dense D2D network, the direct links between the HAP and the receiver can exist in both hops and contribute significantly to the overall throughput. Moreover, leveraging the direct links, we allow the HAP to beamform the same information in two hops. Let $(\mathbf{w}_1, \mathbf{w}_2)$ denote the HAP's signal beamforming strategies in two phases. This can significantly improve the data rate and/or reliability, which is crucial for many applications such as industrial control process.

In the first hop, the HAP beamforms the information with a fixed transmit power p_t and the beamforming vector $\mathbf{w}_1$. Conventionally, the beamforming information can be received by both the relays and the receiver directly, as shown in Fig. 5.1a. Hence, the HAP's beamforming design has to balance between the transmission performance to the relays and to the receiver. We assume that each relay has a dual-mode radio that can switch between the passive and active modes, similar to that in [4, 10]. This arises the novel hybrid relay communications model. As illustrated in Fig. 5.1b, when the HAP beamforms the information signal to the relays, each relay can turn into the passive mode and backscatter the RF signals from the HAP directly to the receiver. By setting a proper load impedance and thus changing the antenna's reflection coefficient [2], the passive relay can backscatter a part of the incident RF signals, while the other part is harvested as the power to sustain its operations. Moreover, the backscattered signals from different passive relays can be coherently combined with the active relays' RF signals to enhance the signal strength at the receiver [11].

The HAP's beamforming in the first hop is also used for wireless power transfer to the relays. We consider a PS protocol for the energy harvesting relays, i.e., a part of the RF signal at the relays is harvested as power while the other part is received as information signal. Specifically, we allow each active relay to set a different PS ratio

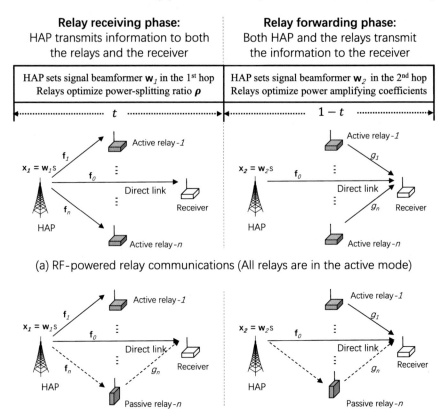

Relay receiving phase:
HAP transmits information to both the relays and the receiver

Relay forwarding phase:
Both HAP and the relays transmit the information to the receiver

HAP sets signal beamformer $\mathbf{w}_1$ in the 1st hop
Relays optimize power-splitting ratio ρ

HAP sets signal beamformer $\mathbf{w}_2$ in the 2nd hop
Relays optimize power amplifying coefficients

$\leftarrow\cdots t \cdots\rightarrow \leftarrow\cdots 1-t \cdots\rightarrow$

(a) RF-powered relay communications (All relays are in the active mode)

(b) Backscatter-assisted hybrid relay communications

Fig. 5.1 Two-hop data transmissions in hybrid relay communications

to match the HAP's beamforming strategy and its energy demand. In the second hop, the active relays amplify and forward the received signals to the receiver. Meanwhile, the HAP also beamforms the same information symbol directly to the receiver with a new beamforming vector $\mathbf{w}_2$. Hence, the received signals at the receiver are a mixture of the signals forwarded by the relays and the direct beamforming from the HAP. With maximal ratio combining (MRC) at the receiver [12], the received signals in both hops can be combined together to enhance the data rate and reliability in transmission.

5.2.2 Channel Enhancement via Passive Relays

The optimal selection of each relay's radio mode is complicated by the relays' couplings in transmission capabilities. The passive mode becomes the only affordable choice when the relay has low power supply, while the active mode can be preferred to provide more reliable transmissions if the relay has good channel conditions and sufficient power supply.

Let $b_n \in \{0, 1\}$ denote the binary variable indicating the radio mode of the relay-n for $n \in \mathcal{N}$. Then the set of relays in Fig. 5.1b will be split into two subsets, i.e., $\mathcal{N}_a \triangleq \{n \in \mathcal{N} : b_n = 0\}$ and $\mathcal{N}_b \triangleq \mathcal{N} \setminus \mathcal{N}_a = \{n \in \mathcal{N} : b_n = 1\}$, denoting the sets of active and passive relays, respectively. Let $\hat{\mathbf{f}}_0$ and $\hat{\mathbf{f}}_k$ for $k \in \mathcal{N}_a$ denote the enhanced channels from the HAP to the receiver and to the active relays, respectively. Let $s(t)$ denote the signal transmitted from the HAP with a constant transmit power p_t. The received signal at the passive relay-n is given by

$$y_n(t) = \sqrt{p_t} \mathbf{f}_n s(t) + v(t), \tag{5.1}$$

where $v(t)$ is the normalized noise signal. The complex reflection coefficient of passive relay-n is given by $\Gamma_n = |\Gamma_n| e^{j\theta_n}$, where $|\Gamma_n|$ denotes the magnitude of signal reflection and $\theta_n \in [0, 2\pi]$ represents the phase offset incurred by backscattering. Hence, the signal received at the receiver is given by

$$d_r = \sqrt{p_t} \mathbf{f}_0 s(t) + \sum_{n \in \mathcal{N}_b} \sqrt{p_t} \mathbf{f}_n s(t) \Gamma_n g_n + \bar{v}(n) = \sqrt{p_t} \hat{\mathbf{f}}_0 s(t) + \bar{v}(n), \tag{5.2}$$

where $\bar{v}(n)$ denotes the aggregate noise signal. Hence, the enhanced channel $\hat{\mathbf{f}}_0$ from the HAP to the receiver can be rewritten as

$$\hat{\mathbf{f}}_0 = \mathbf{f}_0 + \sum_{n \in \mathcal{N}_b} \mathbf{f}_n \Gamma_n g_n = \mathbf{f}_0 + \sum_{n \in \mathcal{N}} b_n \mathbf{f}_n \Gamma_n g_n. \tag{5.3}$$

In the same way, we can rewrite the enhanced channel from the HAP to the active relay-k as

$$\hat{\mathbf{f}}_k = \mathbf{f}_k + \sum_{n \in \mathcal{N}_b} \mathbf{f}_n \Gamma_n z_{n,k} = \mathbf{f}_k + \sum_{n \in \mathcal{N}} b_n \mathbf{f}_n \Gamma_n z_{n,k}, \tag{5.4}$$

where $z_{n,k}$ denotes the complex channel from the passive relay-n to the active relay-k. Note that the enhanced channels $\hat{\mathbf{f}}_0$ and $\hat{\mathbf{f}}_k$ depend not only on the binary indicator b_n, but also the complex reflection coefficient Γ_n of each passive relay in the set $\mathcal{N}_b$. We observe that the phase θ_n is a critical design variable for channel enhancement while $|\Gamma_n|$ can be simply set to its maximum Γ_{max} to increase the reflected signal power.

The channel models in (5.3) and (5.4) are simplified approximations as we omitted the interactions among different passive relays. In fact, besides reflecting the active radios' signals, each passive relay can also reflect the backscattered signals from the other passive relays, thus creating a feedback loop, which makes it very complicated to characterize the the signal models exactly. However, this special case can be reasonable as the double reflections are significantly weakened. In particular, it not only provides the basis for analytical study, but also enables us to derive a lower bound on the throughput performance of hybrid relay communications. A similar channel model has been applied in [11, 13].

5.3 Problem Formulation

In the sequel, we focus on this simplified channel model and formulate the throughput maximization problem for hybrid relay communications, which involves the joint optimization of the HAP's transmit beamforming, the relays' mode selection, and their operating parameters. Assuming a normalized noise power, the receiver's SNR in the first hop is given by

$$\gamma_1 = p_t |\hat{\mathbf{f}}_0^H \mathbf{w}_1|^2, \tag{5.5}$$

where $\hat{\mathbf{f}}_0^H$ is the Hermitian transpose of $\hat{\mathbf{f}}_0$. By controlling the beamformer $\mathbf{w}_1$ in the first hop, the HAP can adjust its information and power transfer to different relays. Each active relay $n \in \mathcal{N}_a$ can choose the different PS ratio ρ_n to balance its power supply and demand:

$$p_n \leq \eta \rho_n p_t |\hat{\mathbf{f}}_n^H \mathbf{w}_1|^2, \quad \forall n \in \mathcal{N}_a, \tag{5.6}$$

where p_n denotes the relay's transmit power and η is the energy harvesting efficiency. The PS ratio ρ_n indicates the portion of RF power converted by the energy harvester. The other part $1 - \rho_n$ is used for signal detection and thus the received signal at the relay-n is given by

$$r_n = \sqrt{(1 - \rho_n) p_t} \hat{\mathbf{f}}_n^H \mathbf{w}_1 s + \sigma_n = y_n s + \sigma_n, \tag{5.7}$$

where we define $y_n \triangleq \sqrt{(1 - \rho_n) p_t} \hat{\mathbf{f}}_n^H \mathbf{w}_1$ for notational convenience and σ_n is the complex Gaussian noise with zero mean and normalized unit variance.

In the second hop, each active relay-n forwards the information to the receiver with the transmit power p_n. All the relays' signals will be combined coherently at the receiver. The power amplifying coefficient of the relay-n is given by $x_n \triangleq \left(\frac{p_n}{1+|y_n|^2}\right)^{1/2}$, which has to be optimized to maximize the overall throughput [14]. Meanwhile, the HAP can transmit the same information symbol directly to the receiver with a new beamformer $\mathbf{w}_2$. This can enhance the reliability and data rate of information transmission from the HAP to the receiver. Hence, the combined signal at the receiver is given by

$$r_d = \sum_{n \in \mathcal{N}_a} x_n \hat{g}_n y_n s + \sum_{n \in \mathcal{N}_a} x_n \hat{g}_n \sigma_n + \sqrt{p_t} \hat{\mathbf{f}}_0^H \mathbf{w}_2 s + v_d. \tag{5.8}$$

The first two terms in (5.8) correspond to the amplified signals by the active relays. The third term represents the HAP's direct beamforming. The channel $\hat{g}_n$ is also an enhanced version of g_n from the relay-n to the receiver. Till this point, we can formulate the SNR in the second hop as

$$\gamma_2 = \frac{\left| \mathbf{x}^T D(\hat{\mathbf{g}}) \mathbf{y} + \sqrt{p_t} \hat{\mathbf{f}}_0^H \mathbf{w}_2 \right|^2}{1 + ||D(\hat{\mathbf{g}}) \mathbf{x}||^2}, \tag{5.9}$$

where $\mathbf{x} = [x_1, x_2, \ldots, x_{|\mathcal{N}_a|}]^T$, $\mathbf{y} = [y_1, y_2, \ldots, y_{|\mathcal{N}_a|}]^T$, and $D(\hat{\mathbf{g}})$ denotes the diagonal matrix with the diagonal element given by $\hat{\mathbf{g}} = [\hat{g}_1, \hat{g}_2, \ldots, \hat{g}_{|\mathcal{N}_a|}]^T$.

Hence, the overall SNR at the receiver can be evaluated as $\gamma = \gamma_1 + \gamma_2$. We aim to maximize SNR in two hops by optimizing the HAP's beamforming strategies $(\mathbf{w}_1, \mathbf{w}_2)$, as well as the relays' radio mode selection b_n and operating parameters, including the PS factor ρ_n and the complex reflection coefficient $\Gamma_n = |\Gamma_n| e^{j\theta_n}$:

$$\max_{\mathbf{w}_1, \mathbf{w}_2, b_n, \rho_n, \theta_n} \quad \gamma_1 + \gamma_2 \tag{5.10a}$$

$$s.t. \quad ||\mathbf{w}_1|| \le 1 \text{ and } ||\mathbf{w}_2|| \le 1, \tag{5.10b}$$

$$b_n \in \{0, 1\}, \quad \forall n \in \mathcal{N}, \tag{5.10c}$$

$$\theta_n \in [0, 2\pi], \quad \forall n \in \mathcal{N}_b. \tag{5.10d}$$

$$\rho_n \in (0, 1), \quad \forall n \in \mathcal{N}_a, \tag{5.10e}$$

$$p_n \le \eta \rho_n p_t |\hat{\mathbf{f}}_n^H \mathbf{w}_1|^2, \quad \forall n \in \mathcal{N}_a. \tag{5.10f}$$

The constraints in (5.10b) denote the HAP's feasible beamforming vectors in two hops. We assume that the HAP's transmit power is fixed at p_t while the beamforming vectors $(\mathbf{w}_1, \mathbf{w}_2)$ can be adapted to maximize the throughput performance. Generally $\mathbf{w}_2$ is not necessarily the same as $\mathbf{w}_1$ as the optimization of $\mathbf{w}_1$ has to take into account the data transmissions to both the relays and the receiver. The binary variable b_n determines the division $(\mathcal{N}_a, \mathcal{N}_b)$ of the relays in different modes. The constraint in (5.10d) ensures that the phase offset of each passive relay in set $\mathcal{N}_b$ is fully controllable via load modulation [4]. Practically, the phase offset θ_n is subject to a finite discrete set. As such, quantization can be applied to obtain the best approximation from the optimal continuous phase offset.

5.4 Performance Maximization with Hybrid Relay Communications

It is obvious that the optimization of the relays' mode selection $(\mathcal{N}_a, \mathcal{N}_b)$ is combinatorial and difficult to solve optimally. Even with fixed relay mode, the throughput maximization is still challenged by the couplings of multiple relays in different modes, which have very different transmission capabilities and energy demands. In the sequel, we solve the throughput maximization problem in a decomposed manner. Firstly, assuming a fixed relay mode, we evaluate the enhanced channels and formulate the throughput maximization with active relays only, similar to that in [10]. Secondly, with the fixed beamforming strategy, we evaluate individual relay's energy status or performance gain, which motivates us to update the relay's mode selection in an iterative manner.

5.4.1 Relay Performance with Fixed Mode Selection

Given a set $\mathcal{N}_b$ of the passive relays and their reflection coefficients Γ_n, the enhanced channels for active RF communications are given by (5.3) and (5.4). Then, we can formulate the throughput maximization problem with the set of active relays alone, which becomes a conventional two-hop relay optimization problem, similar to that in [10, 14]. Our target is to maximize γ by optimizing the HAP's beamforming $(\mathbf{w}_1, \mathbf{w}_2)$ in two hops and the active relays' PS ratios ρ, subject to the relays' power budget constraints:

$$\max_{\mathbf{w}_1,\mathbf{w}_2,\rho_n} \quad \gamma_1 + \gamma_2 \tag{5.11a}$$

$$s.t. \quad ||\mathbf{w}_1|| \leq 1 \text{ and } ||\mathbf{w}_2|| \leq 1, \tag{5.11b}$$

$$\rho_n \in (0, 1), \quad \forall \, n \in \mathcal{N}_a, \tag{5.11c}$$

$$p_n \leq \eta \rho_n p_t |\hat{\mathbf{f}}_n^H \mathbf{w}_1|^2, \quad \forall \, n \in \mathcal{N}_a. \tag{5.11d}$$

Problem (5.11) will achieve different performance when the passive relays set different phase offsets θ_n. The phase optimization can follow the alternating optimization method. In particular, each passive relay initially sets a random phase offset θ_n, based on which we can optimize $(\mathbf{w}_1, \mathbf{w}_2)$ and ρ_n by solving the Problem (5.11). Given the solution to (5.11), we then turn to phase optimization sequentially for each passive relay, which will be detailed in Sect. 5.4.2.

5.4.1.1 Lower Bounds on Relay Performance

In the sequel, we provide a feasible lower bound on (5.11), which is achievable by designing the HAP's beamforming and relaying strategies.

Proposition 5.1 *A feasible lower bound on (5.11) can be found by the convex reformulation as follows:*

$$\max_{\bar{\mathbf{W}}_1, \mathbf{W}_1} \quad p_t||\hat{\mathbf{f}}_0||^2 + p_t\hat{\mathbf{f}}_0^H \mathbf{W}_1\hat{\mathbf{f}}_0 + p_t \sum_{n \in \mathcal{N}_a} s_{n,1} \tag{5.12a}$$

$$s.t. \quad \begin{bmatrix} \kappa_n\psi_n - (1 + \psi_n)s_{n,1} & \sqrt{p_t}s_{n,1} \\ \sqrt{p_t}s_{n,1} & 1 \end{bmatrix} \succeq 0, \quad \forall n \in \mathcal{N}_a, \tag{5.12b}$$

$$\kappa_n \leq \hat{\mathbf{f}}_n^H \mathbf{W}_1\hat{\mathbf{f}}_n, \quad \forall n \in \mathcal{N}_a, \tag{5.12c}$$

$$s_{n,1} = \hat{\mathbf{f}}_n^H \mathbf{W}_1\hat{\mathbf{f}}_n - \hat{\mathbf{f}}_n^H \bar{\mathbf{W}}_1\hat{\mathbf{f}}_n, \quad \forall n \in \mathcal{N}_a, \tag{5.12d}$$

$$\bar{\mathbf{W}}_1 \succeq \mathbf{0} \text{ and } \mathbf{W}_1 \succeq \mathbf{0}. \tag{5.12e}$$

where $\psi_n \triangleq \eta p_t|\hat{g}_n|^2||\hat{\mathbf{f}}_0||^2$ is a constant. At optimum, the PS ratio of the relay-n is given by $\rho_n = \frac{\hat{\mathbf{f}}_n^H \bar{\mathbf{W}}_1\hat{\mathbf{f}}_n}{\hat{\mathbf{f}}_n^H \mathbf{W}_1\hat{\mathbf{f}}_n}$ for $n \in \mathcal{N}_a$.

The proof of Proposition 5.1 follows a similar approach as that in [9], and thus it is omitted here for conciseness. With the fixed relay mode, the channel $\hat{\mathbf{f}}_0$ and $\hat{\mathbf{f}}_n$ can be estimated by a training process. The objective function in (5.12a) then becomes linear and the constraints (5.12b)–(5.12d) define a set of linear matrix inequalities. Hence, Problem (5.12) can be solved efficiently by semidefinite programming (SDP) [15]. Once we find the optimal matrix solution $\mathbf{W}_1$, we can retrieve the HAP's beamforming vector $\mathbf{w}_1$ by eigen-decomposition or Gaussian randomization method [16].

Though exact solution to (5.11) is not available, Problem (5.12) provides a lower bound on the SNR performance, which can serve as the performance metric for the relay's mode selection. It is clear that (5.12c) will hold with equality at optimum and thus we can verify the following property.

Proposition 5.2 *At optimum, the solution to (5.12) is given by $s_{n,1} = (\frac{\rho_n\psi_n}{1-\rho_n} - 1)/p_t$ and the Objective (5.12a) is given by*

$$\gamma = p_t||\hat{\mathbf{f}}_0||^2 + p_t|\hat{\mathbf{f}}_0^H \mathbf{w}_1|^2 + \sum_{n \in \mathcal{N}_a} \left(\frac{\eta\rho_n p_t}{1 - \rho_n}|\hat{g}_n|^2||\hat{\mathbf{f}}_0||^2 - 1 \right). \tag{5.13}$$

The proof of Proposition 5.2 can be referred to [17]. It implies that the lower bound on (5.11) can be evaluated directly from (5.13). However, by an inspection on (5.13), the SNR evaluation can be much less than the optimum of (5.11) if the direct link $\hat{\mathbf{f}}_0$ is practically weak due to physical obstructions. To this end, we also provide another lower bound on (5.11) by ignoring the direct link in the problem formulation. This corresponds to the case when the direct link is blocked or there is a long distance between the transceivers, e.g., [12]. In this case, the objective in (5.11a) is lower bounded by $\frac{|\mathbf{x}^T D(\hat{\mathbf{g}})\mathbf{y}|^2}{1+||D(\hat{\mathbf{g}})\mathbf{x}||^2}$. Let $\hat{x}_n = x_n\hat{g}_n$. The lower bound on (5.11) can be evaluated as

$$\max_{\rho,\hat{\mathbf{x}},\mathbf{y},\mathbf{w}_1} \quad \left|\hat{\mathbf{x}}^T \mathbf{y}\right|^2 (1 + ||\hat{\mathbf{x}}||^2)^{-1} \tag{5.14a}$$

$$s.t. \quad \hat{x}_n^2 \le \bar{x}_n(\rho_n, s_n) \triangleq \frac{\eta \rho_n p_t s_n^2 \hat{g}_n^2}{1 + (1 - \rho_n) p_t s_n^2}, \tag{5.14b}$$

$$y_n^2 \le \bar{y}_n(\rho_n, s_n) \triangleq (1 - \rho_n) p_t s_n^2, \tag{5.14c}$$

$$\rho_n \in (0, 1), \quad \forall\, n \in \mathcal{N}_a \tag{5.14d}$$

$$\hat{\mathbf{x}} \succeq \mathbf{0}, \mathbf{y} \succeq \mathbf{0}, \text{ and } ||\mathbf{w}_1|| \le 1, \tag{5.14e}$$

where $\hat{\mathbf{x}} \triangleq [\hat{x}_1, \hat{x}_2, \dots, \hat{x}_{|\mathcal{N}_a|}]^T$ and we define $s_n^2 = |\hat{\mathbf{f}}_n^H \mathbf{w}_1|^2$ for notational convenience. The upper bounds on $\hat{x}_n^2$ and y_n^2 are defined as $\bar{x}_n(\rho_n, s_n)$ and $\bar{y}_n(\rho_n, s_n)$, respectively, which depend on the relays' PS ratio ρ_n and the HAP's beamforming strategy $\mathbf{w}_1$. Note that $\hat{\mathbf{x}}$ and $\mathbf{y}$ are auxiliary decision variables in Problem (5.14), coupled with the PS ratio ρ and the beamformer $\mathbf{w}_1$ via the constraints in (5.14b) and (5.14c).

5.4.1.2 Alternating Optimization Solution to (5.14)

To the best of our knowledge, there is no exact solution to the non-convex problem in (5.14). Practically, it can be solved by the alternating optimization method that improves the Objective (5.14a) in an iterative manner with guaranteed convergence. In particular, with fixed ρ_n and $\mathbf{w}_1$, the auxiliary variables $\hat{\mathbf{x}}$ and $\mathbf{y}$ are subject to the fixed upper bounds $\bar{x}_n(\rho_n, s_n)$ and $\bar{y}_n(\rho_n, s_n)$, respectively. As such, Problem (5.14) can be viewed as the conventional network beamforming optimization problem with perfect channel information [14], which can be solved optimally in a closed form. However, given the feasible $\hat{\mathbf{x}}$ and $\mathbf{y}$, the optimization of ρ_n and $\mathbf{w}_1$ is still very difficult to solve and thus it requires further approximation within each iteration of the alternating optimization method. To proceed, we first exploit the structural properties of Problem (5.14) that shred some insights on the algorithm design.

Proposition 5.3 *Problem (5.14) has the following properties: (i) $\bar{x}_n(\rho_n, s_n)$ is increasing in both ρ_n and s_n, (ii) $\bar{y}_n(\rho_n, s_n)$ is increasing in s_n and decreasing in ρ_n, and (iii) the constraint in (5.14c) holds with equality at the optimum of (5.14).*

The proof of property (ii) in Proposition 5.3 is straightforward. To verify property (i), we can simply rewrite the upper bound on $\hat{x}_n^2$ as $\bar{x}_n(\rho_n, s_n) = \frac{\eta \rho_n p_t s_n^2 \hat{g}_n^2}{1+(1-\rho_n)p_t s_n^2} = \frac{\eta p_t \hat{g}_n^2}{(1/s_n^2 + p_t)/\rho_n - p_t}$, which is obviously increasing in ρ_n and s_n. Then, we focus on the property (iii). Note that $y_n \ge 0$ only appears in the objective function. If (5.14c) holds with strict inequality at the optimum, we can simply improve the Objective (5.14a) by increasing y_n properly while keeping the other variables unchanged, which brings a contradiction. Therefore, we can guarantee that $y_n = \sqrt{(1 - \rho_n)p_t s_n}$ at the optimum of (5.14).

Algorithm 5.1 Alternating Optimization Solution to (5.14)

1: Initialize ρ_n for $n \in \mathcal{N}_a$, and set $\mathbf{w}_1$ by solving (5.16)
2: $\gamma^{(0)} \leftarrow 0, t \leftarrow 1, \gamma^{(t)} \leftarrow p_t |\hat{\mathbf{f}}_0^H \mathbf{w}_1|^2, \epsilon \leftarrow 10^{-5}, \beta \leftarrow 1/2$
3: **while** $|\gamma^{(t)} - \gamma^{(t-1)}| > \epsilon$
4:     $t \leftarrow t + 1$
5:     Update $\bar{x}_n(\rho_n, s_{\min})$ and $\bar{y}_n(\rho_n, s_{\min})$
6:     Update $\gamma^{(t)}$ and $(\hat{\mathbf{x}}, \mathbf{y})$ by solving Problem (5.15)
7:     Evaluate $G_n(\rho_n, s_{\min})$ for $n \in \mathcal{N}_a$
8:     $m \leftarrow \arg\max_{n \in \mathcal{N}_a} G_n(\rho_n, s_{\min})$
9:     $\rho_m \leftarrow \rho_m - \Delta_m$, where Δ_m is given by (5.17)
10: **end while**

As both $\bar{x}_n(\rho_n, s_n)$ and $\bar{y}_n(\rho_n, s_n)$ are increasing functions of the variable s_n^2, we consider a replacement of s_n by its smallest value $s_{\min}$. As such, we can further derive a lower bound on (5.14) by the following problem:

$$\max_{\rho, \hat{\mathbf{x}} \geq 0, \mathbf{y} \geq 0} \quad |\hat{\mathbf{x}}^T \mathbf{y}|^2 (1 + ||\hat{\mathbf{x}}||^2)^{-1} \tag{5.15a}$$

$$s.t. \quad \hat{x}_n^2 \leq \bar{x}_n(\rho_n, s_{\min}), \tag{5.15b}$$

$$y_n^2 \leq \bar{y}_n(\rho_n, s_{\min}), \tag{5.15c}$$

$$\rho_n \in (0, 1), \quad \forall n \in \mathcal{N}_a. \tag{5.15d}$$

Here $s_{\min}^2$ is given by $s_{\min}^2 = \max_{||\mathbf{w}_1|| \leq 1} \min_{n \in \mathcal{N}_a} |\hat{\mathbf{f}}_n^H \mathbf{w}_1|^2$, which can be easily reformulated into an SDP as follows:

$$\max_{s_{\min}, \mathbf{W}_1} \quad s_{\min}^2 \tag{5.16a}$$

$$s.t. \quad \hat{\mathbf{f}}_n^H \mathbf{W}_1 \hat{\mathbf{f}}_n \geq s_{\min}^2, \quad \forall n \in \mathcal{N}_a, \tag{5.16b}$$

$$\mathbf{W}_1 \succeq \mathbf{0} \text{ and } \text{trace}(\mathbf{W}_1) \leq 1. \tag{5.16c}$$

By solving Problem (5.16) with the interior-point algorithms [15], we can easily determine the HAP's beamformer $\mathbf{w}_1$ from the matrix solution $\mathbf{W}_1$ via Gaussian randomization.

Till this point, we can employ the alternating optimization method to solve (5.15), which provides the lower bound on (5.14) at the convergence. With a fixed PS ratio ρ_n, the feasible solution $(\hat{\mathbf{x}}, \mathbf{y})$ to (5.15) can be easily obtained by the network beamforming optimization in [14], and then we turn to update ρ_n for each active relay based on the following property:

Proposition 5.4 *If the constraint in (5.15b) holds with a strict inequality for some $n \in \mathcal{N}_a$ at optimum, e.g., $\hat{x}_n^2 < \bar{x}_n(\rho_n, s_{\min})$, we can further improve (5.15a) by decreasing ρ_n.*

The proof is straightforward by checking the properties in Proposition 5.3, which reveals that $\bar{x}_n(\rho_n, s_{\min})$ is increasing in ρ_n and y_n is decreasing in ρ_n. At the optimum of (5.15), if $\hat{x}_n^2 < \bar{x}_n(\rho_n, s_{\min})$ for some $n \in \mathcal{N}_a$, we can choose $\Delta_n > 0$ such that $\bar{x}_n^2 \leq \bar{x}_n(\rho_n - \Delta_n, s_{\min}) < \bar{x}_n(\rho_n, s_{\min})$. Meanwhile, we have $y_n^2 = \bar{y}_n(\rho_n, s_{\min}) < \bar{y}_n(\rho_n - \Delta_n, s_{\min})$, which implies that the inequality constraint on y_n can be relaxed. With the relaxed upper bounds on $\hat{x}$ and y, the network beamforming optimization in [14] will produce a higher objective value.

All the above derivations lead to an iterative procedure as shown in Algorithm 5.1. The algorithm starts from a random initialization of ρ_n. The HAP's beamforming strategy w_1 can be obtained by solving (5.16) and kept unchanged during the algorithm iteration. With the fixed (ρ_n, w_1), the upper bounds $(\bar{x}_n, \bar{y}_n)$ are also fixed, and thus we can apply the network beamforming optimization to solve $(\hat{x}, y)$ in Problem (5.15). According to Proposition 5.4, the update of the PS ratio ρ_n can be based on the feasibility check of the inequality constraints in (5.15b) and (5.15c). In particular, for $n \in \mathcal{N}_a$, we search for the relay-n with the largest constraint gap, defined as $G_n(\rho_n, s_{\min}) \triangleq \bar{x}_n(\rho_n, s_{\min}) - \hat{x}_n^2$, and then we reduce ρ_n properly by a small amount Δ_n such that

$$\bar{x}_n(\rho_n - \Delta_n, s_{\min}) = \bar{x}_n(\rho_n, s_{\min}) - \beta G_n(\rho_n, s_{\min}),$$

where $\beta \in (0, 1)$ is a constant parameter. By solving above equation, we can determine Δ_n easily as follows:

$$\Delta_n = \frac{\left(\frac{1}{p_t s_{\min}^2} + 1\right)(\bar{x}_n(\rho_n, s_{\min}) - \beta G_n(\rho_n, s_{\min}))}{\eta + \bar{x}_n(\rho_n, s_{\min}) - \beta G_n(\rho_n, s_{\min})}. \tag{5.17}$$

Once ρ_n is updated as shown in Line 9 of Algorithm 5.1, we turn to optimize $(\hat{x}, y)$ by network beamforming optimization [14].

Moreover, we can show that the lower bound derived by Algorithm 5.1 is also applicable to a single-antenna case. In particular, when the HAP has one single antenna, the enhanced channels $\hat{f}_n$ from the HAP to different relays now become complex variables instead of vectors. The power budget constraints in (5.6) are degenerated to $p_n \leq \eta \rho_n p_t |f_n|^2$. Each active relay's power amplifying coefficient can be similarly defined as $x_n = \left(\frac{p_n}{1+y_n^2}\right)^{1/2}$ where $y_n \triangleq \sqrt{(1 - \rho_n)p_t} f_n$. As such, the performance maximization problem is given by

$$\max_{\rho, \hat{x}, y} \ |\hat{x}^T y|^2 (1 + ||\hat{x}||^2)^{-1} \tag{5.18a}$$

$$\text{s.t. } \hat{x}_n^2 \leq \bar{x}_n(\rho_n, |f_n|) = \frac{\eta \rho_n p_t |f_n|^2 \hat{g}_n^2}{1 + (1 - \rho_n)p_t |f_n|^2}, \tag{5.18b}$$

$$y_n^2 \leq \bar{y}_n(\rho_n, |f_n|) = (1 - \rho_n)p_t |f_n|^2, \tag{5.18c}$$

$$\hat{x} \succeq 0, y \succeq 0, \text{ and } \rho_n \in (0, 1), \quad \forall n \in \mathcal{N}_a. \tag{5.18d}$$

By simply setting $s_n^2 = |\hat{f}_n|^2$, Problem (5.18) has the same form as that in (5.15) and hence it is solvable by Algorithm 5.1.

5.4.2 Iterative Algorithm for Relay Mode Selection

The previous analysis provides the SNR evaluation with the fixed relay mode. This can serve as the performance metric to update each relay's mode selection in an iterative algorithm. The basic idea of the algorithm is to start from a special case with all active relays and then update the relay's radio mode sequentially depending on the relay's performance gain. For simplicity, we allow mode switching of one single relay in each iteration. Hence, the number of iterations will be linearly proportional to the number of relays and the main computational complexity lies in the SNR evaluation given the division $(\mathcal{N}_a, \mathcal{N}_b)$ within each iteration. Such an iterative process continues until no further improvement can be achieved by changing the relays' radio mode.

5.4.2.1 Evaluation of SNR Performance

Considering different channel conditions in the direct link, the SNR evaluation can be performed by solving either the SDP in (5.12) or the non-convex problem in (5.14) by Algorithm 5.1. With the fixed $(\mathbf{w}_1, \boldsymbol{\rho})$, we can evaluate the SNR performance of each relay when it is in the passive mode. After iterating over all relays, we can switch the relay with the maximum SNR performance to the passive mode. It is obvious that the SNR evaluations in (5.12) and (5.14) both rely on the passive relay's complex reflection coefficient $\Gamma_n = |\Gamma_n| e^{j\theta_n}$, which is critical for the channel enhancement in (5.3) and (5.4) and thus effects the SNR evaluation. To maximize the SNR performance, the passive relays can simply set the magnitude of reflection $|\Gamma_n|$ to its maximum $\Gamma_{\max}$. However, the complex phase θ_n is more difficult to optimize due to its couplings cross different relays. The dependence of different channels makes it difficult to enhance all active relays' channels simultaneously.

The optimization of phase shift θ_n can follow the alternating optimization method. In particular, we can firstly optimize the beamforming strategy $(\mathbf{w}_1, \mathbf{w}_2)$ and the relays' PS ratios $\boldsymbol{\rho}$ to maximize the SNR performance. After that, we fix $(\mathbf{w}_1, \mathbf{w}_2)$ and $\boldsymbol{\rho}$, and then turn to optimize the passive relay's reflecting phase $\theta_n \in [0, \pi]$ to further improve the SNR performance. Considering the lower bound in (5.13), the maximum SNR via phase optimization can be approximated as

$$\max_{\theta_n \in [0, \pi]} p_t ||\hat{\mathbf{f}}_0||^2 + p_t |\hat{\mathbf{f}}_0^H \mathbf{w}_1|^2 + \eta p_t ||\hat{\mathbf{f}}_0||^2 g_t, \tag{5.19}$$

where we simply replace $\sum_{n \in \mathcal{N}_a} \frac{\rho_n}{1 - \rho_n} |\hat{g}_n|^2$ in (5.13) by its approximation g_t. This approximation stems from the observation that different active relays are in general spatially distributed with independent channel conditions. As such, the optimization of phase shift θ_n of one passive relay has very limited capability to enhance all the active relays' forwarding channels $\hat{g}_n$ simultaneously. For example, while the forwarding channel $\hat{g}_m$ of an active relay-m is enhanced by the passive relay-n, i.e., $|\hat{g}_m|^2 > |g_m|^2$, another channel $\hat{g}_k$ may become weakened by the same passive relay-n, i.e., $|\hat{g}_k|^2 < |g_k|^2$. Hence, we simply approximate the term $\sum_{n \in \mathcal{N}_a} \frac{\rho_n}{1 - \rho_n} |\hat{g}_n|^2$

Algorithm 5.2 Phase Optimization for SNR Performance

1: Initialize $(\mathbf{w}_1, \rho_n, \Gamma_n)$, $\epsilon \leftarrow 10^{-5}$, $M \leftarrow 20$
2: $\mathrm{SNR}_n^{(0)} \leftarrow 0$, $t \leftarrow 1$, $\mathrm{SNR}_n^{(t)} \leftarrow p_t \|\hat{\mathbf{f}}_0\|^2 + p_t |\hat{\mathbf{f}}_0^H \mathbf{w}_1|^2$
3: **while** $|\mathrm{SNR}_n^{(t)} - \mathrm{SNR}_n^{(t-1)}| > \epsilon$
4: 　　　$t \leftarrow t + 1$
5: 　　　Fix the solution $(\mathbf{w}_1, \rho_n)$
6: 　　　*Phase optimization:*
7: 　　　$\theta_n \leftarrow \arg_{\gamma \in \Theta_d} \max \mathrm{SNR}_n(\gamma)$ by solving (5.20)
8: 　　　$\mathrm{SNR}_n^{(t)} \leftarrow \mathrm{SNR}_n(\theta_n)$
9: 　　　*Beamforming optimization:*
10: 　　　Fix the phase θ_n and update channels by (5.3) and (5.4)
11: 　　　Update $(\mathbf{w}_1, \rho_n)$ by solving (5.12) or (5.14)
12: **end while**
13: **return** Maximum SNR_n and the phase θ_n

---

by a constant $g_t \triangleq \sum_{n \in N_a} \frac{\rho_n}{1-\rho_n} |g_n|^2$. As such, Problem (5.19) can be further rewritten as

$$\max_{\theta_n \in [0,\pi]} \hat{\mathbf{f}}_0^H ((1 + \eta g_t)\mathbf{I} + \mathbf{w}_1 \mathbf{w}_1^H) \hat{\mathbf{f}}_0. \tag{5.20}$$

Let $\mathbf{A} = (1 + \eta g_t)\mathbf{I} + \mathbf{w}_1 \mathbf{w}_1^H \succeq \mathbf{0}$ be the matrix coefficient and $\mathbf{I}$ be the identity matrix. It is clear that Problem (5.20) aims to maximize the compound channel gain $\hat{\mathbf{f}}_0^H \mathbf{A} \hat{\mathbf{f}}_0$ given the HAP's beamforming strategy $\mathbf{w}_1$, which can be easily solved by one-dimension search algorithm.

Let $\mathrm{SNR}_n(\theta_n)$ denote the objective in (5.20) when the relay-n is selected as the passive relay with the reflecting phase θ_n. Assuming a fixed reflection magnitude Γ_{max}, the enhanced channel $\hat{\mathbf{f}}_0$ in (5.3) can be simplified as $\hat{\mathbf{f}}_0 = \mathbf{f}_0 + e^{j\theta_n} \Gamma_{max} g_n \mathbf{f}_n$. The optimal phase θ_n^* can be simply obtained by a one-dimension search method. In particular, we can quantize the continuous feasible region $[0, \pi]$ into a finite discrete set $\Theta_d \triangleq \{0, \pi/M, 2\pi/M, \ldots, \pi\}$, where M denotes the size of Θ_d. Then, we can devise a one-dimension search algorithm to find the phase shift θ_n^* that maximizes $\mathrm{SNR}_n(\theta_n)$. The detailed solution procedure is presented in Algorithm 5.2, which alternates between the HAP's beamforming optimization and the relays' phase optimization by solving two sub-problems in (5.12) and (5.20), respectively. The iteration terminates when $\mathrm{SNR}_n(\theta_n)$ can not be improved anymore.

5.4.2.2 Simplified Schemes for Relay Mode Selection

The convergent value of $\mathrm{SNR}_n(\theta_n)$ in Algorithm 5.2 can be used as a performance metric to evaluate the relay performance when the relay-n is selected as the passive relay. We denote such an SNR-based performance metric as the Max-SNR scheme. However, we note that the SNR evaluation in (5.12) or (5.14) is quite complicated as it requires to solve optimization problems in an iterative procedure. In this part, besides the Max-SNR scheme, we further devise a set of simplified performance metrics based on the problem structure and correspondingly revise the algorithms to update the relays' radio modes.

- Max-Direct-Rate (Max-DR): With a strong direct link from the HAP to the receiver, the data rate contributed by the direct links becomes significant. The Max-DR scheme aims to maximize the data rate via the direct link, which is given by $p_t \|\hat{\mathbf{f}}_0\|^2 + p_t |\hat{\mathbf{f}}_0^H \mathbf{w}_1|^2$. The phase optimization follows a similar approach as that in (5.20), while the beamforming optimization can be significantly simplified comparing to that of the SDP in (5.12).
- Max-Relay-Rate (Max-RR): In contrast to the Max-DR scheme, the Max-RR scheme picks the passive relay that achieves the maximum relay performance without direct links by solving Problem (5.15). This avoids the beamforming optimization in every iteration and thus has significantly reduced complexity than that of (5.12).
- Max-Direct-Gain (Max-DG): The Max-DG scheme further simplifies the optimization in the Max-DR scheme, by focusing on the maximization of the channel gain $|\hat{\mathbf{f}}_0|^2$ instead of the data rate, which can be separated from the beamforming optimization. Hence, the Max-DG scheme avoids the iterative procedure, making it more efficient than the alternating optimization method for solving (5.15).
- Min-RF-Energy (Min-RF): This scheme stems from the intuition that the passive radios are more energy efficient than the active radios. Hence, the relay with low energy supply may prefer to operate in the passive mode. Given the beamforming in (5.12), we can sort the active relays by the RF power at their antennas, i.e., $p_n = \eta \rho_n p_t |\hat{\mathbf{f}}_n^H \mathbf{w}_1|^2$. Thus, the Min-RF scheme switches the active relay with the minimum RF power into the passive mode.

The complete procedure for relay mode selection is presented in Algorithm 5.3. It is initialized with a conventional relay network when all relays are in the active mode. The initial SNR performance SNR_0 serves as the baseline for performance improvement. From Line 6 to Line 9 in Algorithm 5.3, we sequentially select one node from the active relay set $\mathcal{N}_a$ and then evaluate the maximum SNR performance $SNR_n(\theta_n^*)$ when it is turned into the passive mode. If the selected relay can improve the overall SNR performance, we will remove it from the active relay set $\mathcal{N}_a$ and add it to the passive relay set $\mathcal{N}_b$, as shown from Line 10 to Line 15 in Algorithm 5.3. Note that the performance metric for relay mode selection in Line 8 of Algorithm 5.3 can be changed to the Max-DR, Max-RR, Max-DG, and the Min-RF schemes. A performance comparison of different schemes will be presented in Sect. 5.5.

For practical implementation, we assume that the channel conditions vary slowly compared to the runtime of Algorithm 5.3, which is built based on Algorithms 5.1 and 5.2. We also assume that the channel condition of each relay can be estimated sequentially and collected by the HAP in a training process. The main computational task resides in the HAP. Given fixed radio mode, the HAP will call for Algorithm 5.2 to find the optimal operating parameters for each relay. Based on the throughput performance evaluated by Algorithm 5.2, the HAP then updates the relays' radio modes by Algorithm 5.3 and broadcasts the radio mode control message to all relays. In the

---

Algorithm 5.3 Max-SNR based Relay Mode Selection

1: Initialize SNR_0 by solving (5.12) with all active relays
2: $\gamma^{(0)} \leftarrow 0, t \leftarrow 1, \gamma^{(t)} \leftarrow \text{SNR}_0$
3: $\mathcal{N}_a \leftarrow \mathcal{N}, \mathcal{N}_b \leftarrow \emptyset, \epsilon \leftarrow 10^{-5}$
4: **while** $|\gamma^{(t)} > \gamma^{(t-1)}| > \epsilon$
5:     $t \leftarrow t + 1, \gamma^{(t)} \leftarrow \gamma^{(t-1)}$
6:     **for** $n \in \mathcal{N}_a$
7:         Update channels $\hat{\mathbf{f}}_0$ and $\hat{\mathbf{f}}_n$ by (5.3) and (5.4)
8:         Update $\text{SNR}_n(\theta_n^*)$ and $(\mathbf{w}_1, \rho_n)$ by Algorithm 5.2
9:     **end for**
10:     **if** $\max_{n \in \mathcal{N}} \text{SNR}_n(\theta_n^*) > \gamma^{(t-1)}$
11:         $\gamma^{(t)} \leftarrow \max_{n \in \mathcal{N}} \text{SNR}_n(\theta_n^*)$
12:         $n^* \leftarrow \arg\max_{n \in \mathcal{N}} \text{SNR}_n(\theta_n^*)$
13:         $\mathcal{N}_b \leftarrow \mathcal{N}_b \cup \{n^*\}$
14:         $\mathcal{N}_a \leftarrow \mathcal{N}_a - \{n^*\}$
15:     **end if**
16: **end while**
17: **return** $\mathbf{w}_1, (\mathcal{N}_a, \mathcal{N}_b), \{\rho_n\}_{n \in \mathcal{N}_a}$, and $\{\theta_n\}_{n \in \mathcal{N}_b}$

---

next transmission period, the HAP updates the relays' channel conditions and then evaluates the throughput performance again by Algorithm 5.2. The computational complexity of Algorithm 5.3 depends on the number of iterations in the outer loop and that of Algorithm 5.2 in the inner loop. As we only select one passive relay at each iteration, the total number of outer-loop iterations in Algorithm 5.3 will be in the order of N. Within each iteration of Algorithm 5.3, Algorithm 5.2 is called to evaluate the SNR performance. Our numerical results show that the alternating optimization method in Algorithm 5.2 can converge quickly in a few iterations. However, within each iteration we require to solve the SDP in (5.12), which determines the computational complexity of Algorithm 5.2. Given the size of the SDP problem, the computational complexity of (5.12) can be easily characterized by the analytical work in [16], which is a polynomial with respect to the sizes of matrix variables and the number of constraints.

5.5 Numerical Results

In the simulation, we consider the multi-antenna HAP with $K = 3$ antennas and totally $N = 5$ dual-mode energy harvesting relays assisting the information transmission from the HAP to the receiver. To illustrate the radio mode switching of individual relays, we focus on a specific network topology as shown in Fig. 5.2, where the distance in meters between the HAP and the receiver is denoted as $d_0 = 4$. The other relays are spatially distributed between the HAP and the receiver. The location of each relay is also depicted in Fig. 5.2. The conclusions drawn based on a fixed topology may help us understand the connections between channel conditions and the relays' radio mode selections. Throughout the simulations we have the following

Fig. 5.2 Network topology of hybrid relay communications

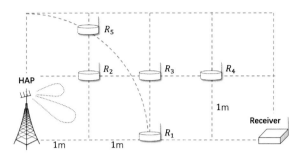

parameter settings unless otherwise stated. The noise power density is -90 dBm and the bandwidth is 100 kHz. The HAP's transmit power p_t in milliwatts can be tuned from 10 to 90. The efficiency for energy harvesting is set to $\eta = 0.5$. The complex modulus of each passive relay's reflection coefficient can be fixed at $\Gamma_{max} = 0.5$. We adopt the log-distance propagation model $L = L_0 + 10\alpha \log_{10}(d/d_0)$, where the loss exponent is $\alpha = 2$ and pass loss in dB at the reference distance $d_0 = 1$ m is $L_0 = 30$. The antenna gain between the transceiver is set to 15 dB, e.g., the HAP can use the directional Antenna ACD24W08 with 12 dBi gain and the user devices use commercial monopole antenna with 3.0 dBi gain. The phase shifts of complex channels are randomly generated. Similar parameter setting can be referred to [10].

5.5.1 Motivation for Hybrid Relay Communications

Intuitively, the relays' radio mode selection depends on individual relays' channel and energy conditions. To verify the performance gain in hybrid relay communications, we consider a three-node relay model with one relay between the HAP and the receiver. By varying the channel conditions in two hops, we examine the relay's optimal radio mode that maximizes the relay performance. Specifically, we fix the channel condition for the first hop between the HAP and the relay, and then vary the phase shift of the complex channel for the second hop from the relay to the receiver. With each fixed channel condition, we evaluate the relay performance with the single relay either in the passive or the active mode. The optimal relay mode is then selected to achieve the maximum relay performance. The numerical result demonstrates that the relay will switch to the passive mode when the phase shifts of the complex channel are set as $4\pi/5$ and $9\pi/5$. This shows some periodicity and verifies the importance of channel conditions on the relay's mode selection. It also implies that the passive relay's phase shift induced by backscattering has to be optimized to match the complex channel conditions. Furthermore, we fix the channel's phases in two hops, and then vary the transmission distance between the HAP and the receiver. Generally, a larger distance implies a higher attenuation to the signal propagation and thus makes the channels worse off. Similarly, we examine the relay performances in two modes and choose the optimal relay mode to maximize the overall throughput. As

Fig. 5.3 The relay's mode selection in a three-node relay model, where one relay locates between the HAP and the receiver. The HAP's transmit power is fixed at $p_0 = 10$ mW. The distance d_0 from the HAP to the receiver increases from 2 to 5 meters

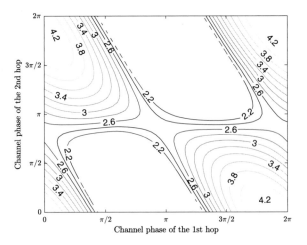

shown in Fig. 5.3, we plot the point of relay's mode switching when we set different transmission distances $d_0 \in (2, 5)$, i.e., the receiver moves far away from the HAP. Each point on the curve in Fig. 5.3 indicates the relay's mode switch from the active to the passive mode. The number attached to each curve represents the distance between the HAP and the receiver at which the relay's mode switching happens. It is obvious that the relay's mode selection is also strongly coupled with the channel gains. In particular, the relay tends to work in the passive mode when the channels become worse off as the distance d_0 becomes large. We also note that there exists some cases in which the relay is always operating in the active mode as the distance d_0 varies.

To verify the throughput performance of our developed algorithm, we also consider a simple two-relay case. Though we cannot find a closed solution to Problem (5.10) even for the two-relay case, we can easily find its optimum by a numerical search method. This allows us to demonstrate the performance gap between our algorithm and the optimum. Specifically, considering the relay-1 and the relay-3 in Fig. 5.2, we have three different combinations for the relays' radio modes, i.e., (relay-1 active, relay-3 passive), (relay-1 passive, relay-3 active), and All-Active. For a fair comparison, we do not consider the All-Passive case as it degenerates to the direct transmission scenario. Once we fix the relays' radio modes, we can adopt the optimal network beamforming algorithm in [14] to jointly optimize the operating parameters of both the HAP and the relays. Then we can compare the throughput performances of these three cases to that of the Max-SNR algorithm. As clearly shown in Fig. 5.4, the maximum performance is achieved when the relay-1 is passive and the relay-2 is active. Most importantly, our Max-SNR algorithm achieves exactly the same as the optimum for the two-relay case.

To verify how passive relays affect the overall throughput performance, we allow all relays in Fig. 5.2 to switch into the passive mode one by one, and then we evaluate the throughput performance in each case. Given the relays' radio modes, we optimize

Fig. 5.4 Performance evaluation for a simple two-relay case

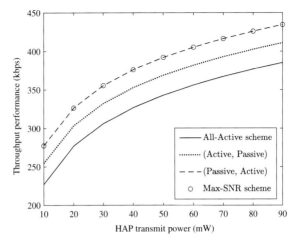

Fig. 5.5 The overall relay performance with a different set of passive relays: 5 active relays are sequentially turned into the passive mode. In baseline case, we have the default setting $(\eta = 0.5, \Gamma = 0.5)$

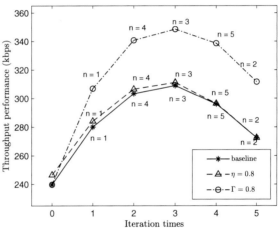

the HAP's beamforming strategy and the relays' operating parameters to maximize the throughput performance. As shown in Fig. 5.5, when we sequentially set the relay-1, relay-4, and relay-3 into the passive mode, the overall throughput performance can be gradually increased. However, the overall throughput becomes worse off when the relay-2 and relay-5 are further turned into the passive mode. The maximum throughput is achieved when both the passive and the active relays collaborate in the relay transmission. In particular, we have $\mathcal{N}_a = \{2, 5\}$ and $\mathcal{N}_b = \{1, 3, 4\}$ for the network in Fig. 5.2. This observation clearly verifies that the developed hybrid relay communications can potentially outperform the conventional relay communications with all relays in the same radio mode.

5.5.2   Comparison of Different Mode Selection Algorithms

For multiple relays in the system, the optimization of individual relays' mode selections becomes combinatorial and problematic. Along with the Max-SNR metric in Algorithm 5.3, we have also developed a set of heuristic algorithms with reduced complexity. The Max-SNR scheme evaluates a lower bound of the relay performance and relies on an iterative procedure to improve the relay performance by updating the relays' mode selection in each iteration. A few heuristic algorithms are also developed to simplify the evaluation of the overall SNR performance. The Max-DR scheme focuses on the throughput maximization of the direct link from the HAP to the receiver, while the Max-RR scheme maximizes the relay performance without the direct links. The Max-DG scheme avoids the iterative procedure by simply maximizing the channel gain of the direct links. In the last, the Min-RF scheme avoids the trouble in solving optimization problems and updates the relays' mode selections based on an ordering of their RF energy at the antennas.

Figure 5.6 shows the overall throughput performance by using different algorithms for the relays' mode selection. We find that the Max-SNR scheme significantly outperforms the other schemes, however with the cost of a higher computational complexity. Note that both of the Max-DR and the Max-RR schemes are relying on simplified approximations of the overall SNR performance. In Fig. 5.6, we observe that the Max-DR scheme achieves higher performance than that of the Max-RR scheme, which implies that the direct links contribute significantly to the overall relay performance in the evaluated network topology. An interesting observation is that the most simple Max-DG scheme even outperforms that of the Max-DR scheme, which requires an iterative procedure to jointly optimize the reflecting phase and the HAP's beamforming strategy. Therefore, we can conclude that the Max-DG scheme is practically good in terms of throughput performance and computational complexity. The Max-SNR scheme can be viewed as a performance benchmark. In the sequel,

Fig. 5.6 Comparison of different mode selection algorithms

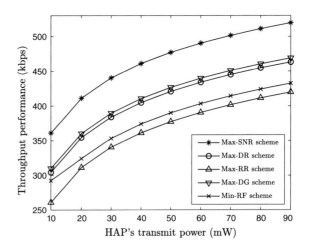

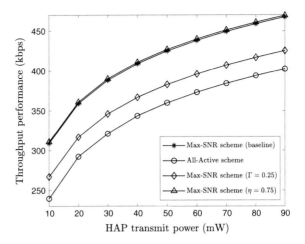

Fig. 5.7 Throughput performance of the Max-SNR algorithm

we focus on the Max-SNR scheme and verify its performance dynamics with respect to different parameters.

5.5.3 Throughput Dyanmics in the Max-SNR Algorithm

In Fig. 5.7, we evaluate the throughput performance of the Max-SNR scheme with different transmit power at the HAP. For comparison, we also show the relay performance when all relays are working in the active mode, which is denoted as the All-Active scheme in Fig. 5.7. In the baseline algorithm, we set $\Gamma = 0.5$ and $\eta = 0.5$. These parameters are also varied to examine the performance of the Max-SNR algorithm when the relays have different capabilities in signal reflection and energy harvesting. The most straightforward observation is that the Max-SNR algorithm achieves the highest throughput compared to the All-Active scheme, which coincides with our observation in Fig. 5.5 and corroborates our motivation to optimize the relays' mode selection in hybrid relay communications. In particular, with 50 mW transmit power at the HAP, the Max-SNR algorithm achieves nearly 20 % throughput improvement comparing to that of the conventional All-Active scheme. Moreover, we observe that the overall relay performance barely increases when the relays can harvest RF energy more efficiently with a larger value $\eta = 0.75$ comparing to the baseline $\eta = 0.5$. This can be understood by the extremely low power consumption of the passive relays. Even with a smaller energy harvesting efficiency, the RF power can be still sufficient to power the passive relays' operations. As such, the overall relay performance becomes insensitive to the relays' energy harvesting, however it can be very sensitive to the passive relays' capabilities of signal reflections. As demonstrated in Fig. 5.7, the throughput is decreased significantly when we set a smaller value $\Gamma = 0.25$ comparing to the baseline $\Gamma = 0.5$.

Fig. 5.8 The relay's mode selection improves the overall throughput

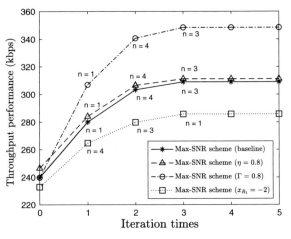

Fig. 5.9 The update of relay mode happens in the outer loop. Within each loop, the algorithm optimizes $\mathbf{w}_1$ and ρ iteratively

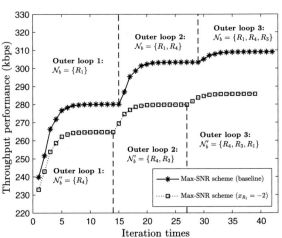

In this part, we intend to show how each relay changes its radio mode as the algorithm iterates. Figure 5.8 shows the throughput dynamics of the Max-SNR based Algorithm 5.3 in each iteration. We also show the index of the passive relay on the throughput curve when it is switched into the passive mode. As illustrated in Fig. 5.8, the Max-SNR algorithm initializes all relays in the active mode. As the algorithm iterates, the relay-1, relay-4, and relay-3 are sequentially switched to the passive mode to improve the overall relay performance. The throughput dynamics in the inner loop of Algorithm 5.3 are detailed in Fig. 5.9. Each outer loop implies the update of one relay's radio mode, while each inner loop indicates an iteration in the alternating optimization of $\mathbf{w}_1$ and ρ. With the fixed relay mode, we can observe that the throughput is improved in each iteration of the alternating optimization method. Algorithm 5.3 terminates when the throughput performance cannot be further improved by changing the relays' radio mode. Note that the relay-1, relay-4, and relay-3 are generally far away from the HAP while closer to the receiver, as observed from the network topology in Fig. 5.2. Hence, the results in Fig. 5.8 imply that the relays with worse channel conditions in the first hop may prefer to operate in the passive mode. With different parameters $\Gamma = 0.8$ and $\eta = 0.3$, the results in Fig. 5.8 verify that the throughput performance is insensitive to the relays' capabilities of energy harvesting but more sensitive to the capabilities of signal reflections. Figure 5.8 also shows the throughput dynamics when the relay-1 changes its location (i.e., the x-coordinate of the relay-1 is shifted to -2), which implies the change of the relays' channel conditions and thus affects the sequence of radio mode switching in Algorithm 5.3.

5.6 Conclusions

In this chapter, we have introduced the novel concept of hybrid relay communications involving both the active and passive relays, and then presented a throughput maximization problem by jointly optimizing the HAP's beamforming, the relays' mode selection and operating parameters. Two lower bounds are derived to evaluate the SNR performance under different channel conditions, which further serve as the performance metric for updating the relays' mode selection in an iterative manner. We have also devised a few heuristic performance metrics for the relays' mode selection with much reduced computational complexity. Simulation results have verified the advantageous of using hybrid relay communications, and demonstrated that a significant performance gain can be achieved by the developed algorithms comparing to the conventional relay communications with all active relays.

References

1. V. Liu, A.N. Parks, V. Talla, S. Gollakota, D. Wetherall, J.R. Smith, Ambient backscatter: wireless communication out of thin air, in *ACM SIGGOMM* (ACM, 2013), pp. 39–50
2. N.V. Huynh, D.T. Hoang, X. Lu, D. Niyato, P. Wang, D.I. Kim, Ambient backscatter communications: a contemporary survey. IEEE Commun. Surveys Tuts. **20**(4), 2889–2922 (2018)
3. D.T. Hoang, D. Niyato, P. Wang, D.I. Kim, Z. Han, Ambient backscatter: a new approach to improve network performance for RF-powered cognitive radio networks. IEEE Trans. Commun. **65**(9), 3659–3674 (2017)
4. S. Gong, J. Xu, D. Niyato, X. Huang, Z. Han, Backscatter-aided cooperative relay communications in wireless-powered hybrid radio networks. IEEE Netw. **33**(5), 234–241 (2019)
5. X. Lu, D. Niyato, H. Jiang, E. Hossain, P. Wang, Ambient backscatter-assisted wireless-powered relaying. IEEE Trans. Green Commun. Netw. **3**(4), 1087–1105 (2019)
6. S. Gong, L. Gao, J. Xu, Y. Guo, D.T. Hoang, D. Niyato, Exploiting backscatter-aided relay communications with hybrid access model in device-to-device networks. IEEE Trans. Cogn. Commun. Netw. **5**(4), 835–848 (2019)
7. B. Lyu, Z. Yang, T. Xie, G. Gui, F. Adachi, Optimal time allocation in relay assisted backscatter communication systems, in *IEEE VTC Spring* (IEEE, 2018), pp. 1–5

8. A.A. Nasir, X. Zhou, S. Durrani, R.A. Kennedy, Relaying protocols for wireless energy harvesting and information processing. IEEE Trans. Wireless Commun. **12**(7), 3622–3636 (2013)
9. X. Luo, J. Xu, Y. Zou, S. Gong, L. Gao, D. Niyato, Collaborative relay beamforming with direct links in wireless powered communications, in *IEEE WCNC* (IEEE, 2019), pp. 1–6
10. J. Xu, J. Li, S. Gong, K. Zhu, D. Niyato, Passive relaying game for wireless powered internet of things in backscatter-aided hybrid radio networks. IEEE Internet Things J. **6**(5), 8933–8944 (2019)
11. G. Yang, Q. Zhang, Y.-C. Liang, Cooperative ambient backscatter communications for green Internet-of-Things. IEEE Internet Things J. **5**(2), 1116–1130 (2018)
12. Y. Liu, Wireless information and power transfer for multirelay-assisted cooperative communication. IEEE Commun. Lett. **20**(4), 784–787 (2016)
13. R. Long, Y.-C. Liang, H. Guo, G. Yang, R. Zhang, Symbiotic radio: a new communication paradigm for passive internet of things. IEEE Internet Things J. **7**(2), 1350–1363 (2020)
14. Y. Jing, H. Jafarkhani, Network beamforming using relays with perfect channel information. IEEE Trans. Inf. Theory **55**(6), 2499–2517 (2009)
15. L. Vandenberghe, S. Boyd, Semidefinite programming. SIAM Rev. **38**(1), 49–95 (1996)
16. Z.-Q. Luo, W.-K. Ma, A.-C. So, Y. Ye, S. Zhang, Semidefinite relaxation of quadratic optimization problems. IEEE Signal Process. Mag. **27**(3), 20–34 (2010)
17. Y. Xie, Z. Xu, S. Gong, J. Xu, H.T. Dinh, D. Niyato, Backscatter-assisted hybrid relaying strategy for wireless powered IoT communications, in *IEEE Globecom* (IEEE, 2019), pp. 1–6

Chapter 6
Summary

In this chapter, we summarize this book and discuss the future directions for the resource allocation in backscatter-assisted communication networks.

6.1 Summary of Contributions

In this book, we mainly investigate the resource allocation in backscatter-assisted communication networks, where the performance tradeoff between backscatter communications and traditional communications is particularly considered. Specifically, for backscatter-assisted RF-powered CR networks, we mainly investigate the auction-based time scheduling, contract-based time assignment, and the evolutionary game-based AP and service selection. For backscatter-assisted hybrid relay networks, we mainly investigate the throughput-maximized relay mode selection and resource sharing. The main contributions of this book are summarized as follows.

1. In Chap. 2, we investigate the auction-based time scheduling for backscatter-assisted RF-powered CR networks. With many STs connected to the network, the total transmission demand of the STs may frequently exceed the transmission capacity of the secondary network. Therefore, according to a variety of demand requirements from STs, we design two auction-based time scheduling mechanisms for the time resource assignment. In the auctions, the SG acts as the seller as well as the auctioneer, and the STs act as the buyers to bid for the time resource. We design the winner determination, the time scheduling, and the pricing schemes for both the developed auction-based mechanisms. Furthermore, the economic properties such as IR and truthfulness, and the computational efficiency of our developed mechanisms are analytically evaluated. Simulation results demonstrate the effectiveness of our developed mechanisms.

© The Author(s), under exclusive license to Springer Nature Singapore Pte Ltd. 2021
X. Gao et al., *Resource Allocation in Backscatter-Assisted Communication Networks*,
https://doi.org/10.1007/978-981-16-5127-4_6

2. In Chap. 3, we consider a practical scenario, where the SG only knows the statistical information about the harvested power of the ST, and develop a contract-based time assignment scheme for the backscatter-assisted RF-powered CR networks. Specifically, the SG designs a contract, including a series of time-price items, to maximize its profit. Then, the ST accepts the contract item which can maximize its utility. We derive the optimal contract, which satisfies the IC and the IR properties. Numerical results are presented to verify the effectiveness of our designed contract for the time assignment in backscatter-assisted RF-powered CR network.

3. In Chap. 4, we investigate the evolutionary game-based dynamic AP and service selection in a backscatter-assisted RF-powered CR network, where many STs can choose different transmission services provided by multiple APs. Specifically, we model the AP and service adaptation of the STs by the replicator dynamics, and analytically prove the existence and uniqueness, and the stability of the evolutionary equilibrium. We also consider the delay of information used by the STs to adapt their selection and perform the analysis by using delayed replicator dynamics. In particular, the stability region of the delayed replicator dynamics in a special case is derived. Furthermore, we develop a low-complexity algorithm for the AP and service selection in the network based on evolutionary game. Extensive simulations have been conducted to demonstrate the effectiveness of the developed AP and service selection strategy in the network.

4. In Chap. 5, we investigate the throughput-maximized relay mode selection and resource sharing in backscatter-assisted hybrid relay networks. We aim to jointly optimize the HAP's beamforming, individual relays' radio modes, PS ratios, and the relays' collaborative beamforming strategies to enhance the throughput performance at the receiver. The resulting formulation becomes a combinatorial and non-convex problem. We firstly develop a convex approximation to the original problem, which serves as a lower bound of the relay performance. Then, we design an iterative algorithm that decomposes the binary relay mode optimization from the other operating parameters. In the inner loop of the algorithm, we exploit the structural properties to optimize the relay performance with the fixed relay mode by using alternating optimization. In the outer loop, different performance metrics are derived to guide the search for a set of passive relays to further improve the relay performance. Simulation results verify the effectiveness of the developed scheme in terms of improving the network throughput.

6.2 Future Directions

In this section, we mainly discuss the future directions for the resource allocation in backscatter-assisted communication networks.

6.2.1 Multi-objective Resource Allocation in Backscatter-Assisted Communication Networks

The resource allocation schemes for backscatter-assisted communication networks in this book are mainly to improve the throughput or the amount of the transmitted data. Actually, there are many other metrics which should be also paid attention to. For example, in some scenarios, where the transmission delay is of special importance. Therefore, considering the characteristics of backscatter-assisted communication networks and the development trend of the communication devices, investigating the tradeoff among the performance metrics, such as throughput and delay, and developing the resource allocation schemes considering multiple performance metrics is essential.

6.2.2 Robust Resource Allocation in Backscatter-Assisted Communication Networks

The resource allocation schemes for backscatter-assisted communication networks in this book are mainly investigated based on perfect channel state information. However, due to the channel estimation errors, feedback delay, and channel variation over time, perfect channel state information is challenging to obtain. Therefore, extending our developed resource allocation schemes to the scenario with imperfect channel state information would be more practical, and it is therefore worth to be investigated.

6.2.3 Incentive Resource Allocation in Backscatter-Assisted Mobile Communication Networks

The resource allocation schemes for backscatter-assisted communication networks in this book are mainly developed based on the fixed topology. With the rapid development of unmanned aerial vehicles (UAVs), developing the incentive resource allocation schemes with the mobile topology can provide more insights for the future communication networks. Therefore, considering some incentive properties, such as IC and IR, investigating the incentive resource allocation schemes for backscatter-assisted mobile communication networks is worth to be investigated.

6.2.4 Incentive Resource Allocation in Backscatter-Assisted Hybrid Relay Networks

With more and more devices connected to the network, developing incentive resource allocation schemes is very important. Game theory is powerful to develop incentive resource allocation schemes and analyze distributed networks. Therefore, based on game theory, such as Stackelberg game and auction, taking the characteristics of backscatter-assisted hybrid relay networks, developing incentive resource allocation schemes for hybrid relay networks is important, which can provide useful guidance for the design of future backscatter-assisted hybrid relay networks.

6.2.5 Incentive Resource Allocation in Backscatter-Assisted Communication Networks with Non-orthogonal Multiple Access (NOMA)

NOMA is an effective way to coordinate the access of multiple devices. Therefore, considering the characteristics of backscatter-assisted communication networks and the development trend of communication devices, developing resource allocation schemes in backscatter-assisted communication networks with NOMA is important. Note that in the resource allocation schemes for the networks with NOMA, some issues, such as the information asymmetry and bounded rationality of users, also need to be considered. Therefore, developing incentive resource allocation scheme for backscatter-assisted communication networks with NOMA is worth to be investigated.

Printed in the United States
by Baker & Taylor Publisher Services